New Technologies in Agronomy for Sustainability

Janardan Singh
CSK Himachal Pradesh Krishi Vishvavidalaya, India

Shilpa
Lovely Professional University, India

Pooja
CSK Himachal Pradesh Krishi Vishvavidalaya, India

IGI Global
Scientific Publishing
Publishing Tomorrow's Research Today

Vice President of Editorial	Melissa Wagner
Director of Acquisitions	Mikaela Felty
Director of Book Development	Jocelynn Hessler
Production Manager	Mike Brehm
Cover Design	Jose Rosado

Published in the United States of America by
IGI Global Scientific Publishing
701 East Chocolate Avenue
Hershey, PA, 17033, USA
Tel: 717-533-8845 | Fax: 717-533-7115
Website: https://www.igi-global.com E-mail: cust@igi-global.com

Library of Congress Cataloging-in-Publication Data

Names: Singh, Janardan, editor. | Manhas, Shilpa, 1995- editor. | Jhorar, Pooja, 1998- editor.
Title: New technologies in agronomy for sustainability / edited by Janardan Singh, Shilpa, Pooja.
Description: Hershey, PA : IGI Global, [2026] | Includes bibliographical references and index. | Summary: "The book aims to identify and explore cutting-edge technologies that have the potential to revolutionize agronomy. By providing insights into innovations such as precision agriculture, biotechnology, digital farming, and others, it seeks to showcase their capabilities and applications in advancing sustainability within agricultural systems"-- Provided by publisher.
Identifiers: LCCN 2024038911 (print) | LCCN 2024038912 (ebook) | ISBN 9798369343920 (hardcover) | ISBN 9798369343968 (paperback) | ISBN 9798369343937 (ebook)
Subjects: LCSH: Agricultural innovations. | Precision farming. | Artificial intelligence--Agricultural applications. | Plant biotechnology.
Classification: LCC S494.5.I5 N494 2024 (print) | LCC S494.5.I5 (ebook) | DDC 630.2/084--dc23/eng/20240925
LC record available at https://lccn.loc.gov/2024038911
LC ebook record available at https://lccn.loc.gov/2024038912

British Cataloguing in Publication Data
A Cataloguing in Publication record for this book is available from the British Library.

Table of Contents

Detailed Table of Contents

Climate change poses a major environmental challenge of the 21st century, causing long-term shifts in climate patterns that affect agriculture. This includes changes in temperature, rainfall, and carbon dioxide levels, which impact plant growth, crop yields, and food security. In India, where agriculture is vital to the economy and employs a large workforce, climate change exacerbates issues like rainfall dependence and crop profitability. To combat these challenges, India must integrate adaptation and mitigation strategies, including climate-smart agriculture and enhanced weather data. Collaboration among policymakers, researchers, and farmers is crucial to improve resilience and ensure sustainable food production and economic security.

As the global population approaches 9.7 billion by 2050, sectors such as agriculture and water supply are facing unprecedented challenges. To address these pressures, enhancing resource resilience and sustainability is crucial. Technological innovations in AI and ICT offer a promising solution. Tools like Conserwater, SupPlant, Plantix, and advanced smart irrigation systems are revolutionizing how we manage resources. These technologies provide real-time monitoring and decision support, optimizing efficiency. By leveraging AI techniques such as machine learning and deep learning, integrated with GIS, smart meters, and mobile devices, we gain predictive insights

increased by 44% and cucumber by 33%, while the quality of the products met the appropriate requirements.

T. Kumaresan, Department of Mechanical Engineering, PSG Polytechnic College, Coimbatore, India
S Prabhu, Department of Mechatronics Engineering, PSG Polytechnic College, Coimbatore, India
Avni Suresh Thakkar, School of Management, Shri Ramdeobaba University of Engineering and Management, Nagpur, India
Sampada Viraj Dravid, Department of Mechanical Engineering, Vishwakarma Institute of Technology, Pune, India
V. Krishnamoorthy, Department of Management Studies, Kongu Engineering College, Perundurai, India

Additional technologies commonly referred to as 3D and 4D printing, is emerging as a transformative technology in sustainable agriculture in the structure of the circular economy. Local of agricultural equipment, parts of equipment, and local, enabling production on demand, reduces AM waste, reduces carbon emissions and increases resource efficiency. This chapter explores how innovative AM technologies support clause-loop agricultural systems by reducing fake, customization and supply chain dependence. It also discusses the integration of renewable and bio-based materials in AM processes, promoting regenerative methods. Eventually, AM provides a scalable way to elastic, eco-efficient and circular agricultural value chains for the future.

B. M. Parashiva Murthy, Department of Computer Science and Engineering, JSS Science and Technology University, Mysuru, India
Rajesh V. Argiddi, Department of Computer Science and Engineering, Walchand Institute of Technology, Solapur, India
Gunti Bhupal Raj, School of Agriculture, SR University, Warangal, India
M. Prasanth, Nandha Engineering College, Perundurai, India
Venkateswara Rao Podile, KL Business School, Koneru Lakshmaih Education Foundation, Vaddeswaram, India
M. Suresh, KPR Institute of Technology, Kovai, India

Recent progresses in genetic engineering and biotechnology are revolutionizing crop yields, resistance to pests and diseases and environmental stability. Techniques such as CRISPR-KAS9, transgenic modification and molecular marker-support selection are enabling specific feature improvement and rapid breeding cycle. These innovations contribute to climate-elastic agriculture, reduction of agrochemical use and improved crop nutrient profiles. Moreover, biotechnological programs are supporting land health and sustainable farming practices. This chapter detects current progresses, practical applications and future possibilities in agronomies that highlight their transformative impact on global food security and agricultural productivity.

Preface

INTRODUCTION

In an era marked by rapid population growth, climate variability, and increasing pressure on finite natural resources, agriculture finds itself at a decisive juncture. The responsibility of feeding a growing global population while preserving environmental integrity and social well-being has never been more complex—or more urgent. As Editors, we are pleased to present *New Technologies in Agronomy for Sustainability*, a reference volume that responds to this challenge by bringing together contemporary scientific knowledge, technological innovation, and practical insight aimed at shaping a more sustainable and resilient agricultural future.

This edited volume has been conceived as both a scholarly resource and a practical guide. It captures the collective expertise of leading researchers, academicians, and practitioners who are actively engaged in advancing agronomic science through innovative technologies. The chapters explore a wide spectrum of emerging and transformative approaches, including precision agriculture, speed breeding, digital and smart farming systems, biotechnology, and organic and resource-efficient practices. Together, these contributions demonstrate how modern technologies can be strategically integrated to enhance productivity, optimize resource use, and mitigate the environmental impacts of agricultural intensification.

At the core of this book lies a strong commitment to sustainability as a guiding principle. We adopt a holistic perspective that recognizes the interdependence of ecological health, economic viability, and social equity.

Rather than viewing technology as an end in itself, the chapters emphasize its role as an enabler—supporting resilient farming systems, conserving soil and water resources, protecting biodiversity, and strengthening food and nutritional security. In doing so, this volume advocates a paradigm shift in agronomy, one that moves beyond short-term gains toward long-term sustainability and stewardship of natural resources.

The objectives of this book are multifaceted. First, it seeks to identify and critically examine emerging technologies with the potential to transform agronomic practices. By elucidating their principles, applications, and limitations, the book provides readers with a clear understanding of how these innovations can be deployed in diverse agro-ecological contexts. Second, it promotes sustainable agricultural practices by illustrating how technological interventions can improve efficiency, reduce environmental footprints, and enhance system resilience. Third, the book directly addresses some of the most pressing challenges facing global agriculture today, including food insecurity, soil degradation, water scarcity, climate change, and biodiversity loss, offering science-based and technology-driven solutions.

Equally important is the book's role as a platform for knowledge sharing and collaboration. By bringing together interdisciplinary perspectives, it encourages dialogue among researchers, educators, extension professionals, policymakers, and industry stakeholders. The inclusion of empirical studies, case examples, and expert analyses aims to support evidence-based decision-making and to inform policy development at local, national, and global levels. We hope that these insights will assist policymakers and development practitioners in designing strategies that foster sustainable agricultural transformation.

This volume is intended for a broad and diverse audience. Agronomists, agricultural engineers, scientists, students, and educators will find it valuable as a reference that synthesizes current research and emerging trends. Policymakers and development professionals may draw upon its analyses to support informed planning and governance. Farmers and agricultural practitioners, particularly those interested in adopting advanced technologies such as precision farming, speed breeding, and data-driven decision tools, will find practical knowledge to enhance productivity while maintaining environmental responsibility.

As Editors, we express our sincere gratitude to all contributing authors for their scholarly rigor, dedication, and willingness to share their expertise. We also acknowledge the efforts of reviewers and the support of the publisher in bringing this volume to fruition. It is our hope that *New Technologies in Agronomy for Sustainability* will serve as a valuable reference, stimulate further research and innovation, and contribute meaningfully to the global pursuit of a just, resilient, and sustainable agricultural system for present and future generations.

ORGANIZATION OF THE BOOK

Chapter 1 addresses one of the most pressing challenges confronting global agriculture—climate change. This chapter provides a focused examination of both mitigation and adaptation strategies, with particular emphasis on climate-smart agriculture and the role of improved weather intelligence. Drawing on the Indian agricultural context, it underscores the vulnerability of climate-dependent farming systems and the socioeconomic implications for food security and rural livelihoods. As Editors, we view this chapter as a critical contribution that emphasizes collaboration among policymakers, researchers, and farming communities to strengthen resilience and ensure sustainable agricultural development under changing climatic conditions.

Chapter 2 examines the growing role of Artificial Intelligence (AI) and Information and Communication Technology (ICT) in efficient nutrient and water management. In the context of a rapidly expanding global population and increasing resource scarcity, this chapter demonstrates how AI-driven tools and smart irrigation systems enable real-time monitoring, predictive analytics, and decision support. By integrating machine learning, GIS, smart sensors, and mobile platforms, the chapter illustrates a technological shift toward precision resource management. We view this contribution as a compelling illustration of how digital innovation can enhance sustainability, resilience, and productivity in agricultural systems.

Chapter 3 provides a comprehensive review of Artificial Intelligence applications in smart agriculture. The chapter discusses how AI, machine learning, and deep learning techniques are being used to manage critical agricultural variables such as soil conditions, plant health, weather variability, and yield prediction. By addressing uncertainty in farming environments, the chapter demonstrates how AI-supported decision-making can improve efficiency, reduce costs, and enable "more production with fewer resources." We consider this chapter an essential synthesis that showcases AI as a cornerstone of next-generation agricultural systems.

Chapter 4 addresses the global challenge of problematic soils and waters, which threaten agricultural productivity and food security amid declining arable land. This chapter offers an in-depth analysis of soil degradation caused by salinity, sodicity, acidification, and unsustainable farming practices. It highlights recent advances in soil remediation, including biochar application, precision fertilization, water-saving irrigation, and biological reclamation using salt-tolerant species. As Editors, we value this chapter for its practical and science-based strategies that demonstrate how degraded soils can be restored into productive resources through sustainable management approaches.

Chapter 5 presents empirical evidence on the use of electrochemically activated water to enhance potato yield and storage safety under Western Siberian conditions. Based on multi-year field trials, the chapter demonstrates significant yield improvements and reduced disease incidence through the use of catholyte and anolyte treatments. The environmentally friendly nature of this approach aligns strongly with the sustainability theme of the book. We consider this chapter a valuable example of how innovative, low-impact technologies can improve both productivity and post-harvest quality.

Chapter 6 examines pre-sowing seed treatment using activated water as an environmentally friendly growth stimulant for vegetable crops. Through industrial-scale testing on tomatoes and cucumbers, the chapter reports substantial yield increases while maintaining product quality standards. As Editors, we highlight this chapter as a practical and scalable solution that underscores the potential of simple yet innovative technologies to enhance crop performance. It reinforces the book's overarching message that sustainability-driven innovations can be both effective and accessible for widespread agricultural adoption.

Chapter 7 explores how innovative AM technologies support clause-loop agricultural systems by reducing fake, customization and supply chain dependence. It also discusses the integration of renewable and bio-based materials in AM processes, promoting regenerative methods. Eventually, AM provides a scalable way to elastic, eco-efficient and circular agricultural value chains for the future.

Chapter 8 concludes the volume by detecting current progresses, practical applications and future possibilities in agronomies that highlight their transformative impact on global food security and agricultural productivity.

CONCLUSION

As Editors, we conclude this volume with a renewed sense of optimism and responsibility for the future of agronomy and global food systems. *New Technologies in Agronomy for Sustainability* brings together a diverse yet interconnected body of knowledge that reflects the rapid evolution of agricultural science in response to unprecedented environmental, climatic, and socio-economic challenges. The chapters collectively demonstrate that sustainability in agriculture is no longer an aspirational ideal, but an achievable outcome when innovation, scientific rigor, and responsible management converge.

Throughout this book, emerging technologies—from circular economy–based management practices and climate-smart strategies to artificial intelligence, machine learning, biotechnology, precision tools, and environmentally friendly growth stimulants—have been shown to play a transformative role in enhancing productivity while conserving natural resources. These contributions emphasize that technological advancement must be guided by ecological sensitivity, ethical considerations, and inclusivity to ensure long-term resilience of agricultural systems. The integration of digital intelligence, biological innovation, and sustainable resource management presented in this volume illustrates a clear pathway toward reducing environmental footprints while strengthening food and nutritional security.

Importantly, this book underscores the value of interdisciplinary collaboration and knowledge exchange among researchers, practitioners, policymakers, and farmers. Sustainable agronomy cannot be achieved in isolation; it requires coordinated efforts that bridge scientific discovery with field-level implementation and supportive policy frameworks. The case studies, empirical findings, and strategic insights presented across chapters offer evidence-based guidance that can inform decision-making at local, national, and global scales.

As we look ahead, we believe this volume will serve not only as a reference for current research and practice but also as a catalyst for future innovation. The challenges confronting agriculture will continue to evolve, but so too will the tools and strategies available to address them. We hope that this book inspires continued inquiry, responsible adoption of new technologies, and a shared commitment to sustainability among all stakeholders in the agricultural sector.

In closing, *New Technologies in Agronomy for Sustainability* reaffirms the critical role of science-driven innovation in shaping resilient, productive, and environmentally sound agricultural systems. It is our sincere hope that the insights presented herein will contribute meaningfully to building a more sustainable and equitable agricultural future for generations to come.

Chapter 1

Climate Change Mitigation and Adaptation Strategies in Agriculture

Pankaj Dahiya
https://orcid.org/0000-0002-7625-5451
Lovely Professional University, Punjab, India & CCS Haryana Agriculatural University, Hisar, India

Sachin Kumar
https://orcid.org/0009-0008-4604-8700
CCS Haryana Agriculatural University, Hisar, India

Muskan Yadav
https://orcid.org/0009-0008-7180-8249
CCS Haryana Agriculatural University, Hisar, India

Pooja
https://orcid.org/0009-0007-1120-5975
CCS Haryana Agriculatural University, Hisar, India

Anil
https://orcid.org/0009-0009-9898-7563

DOI: 10.4018/979-8-3693-4392-0.ch001

CCS Haryana Agriculatural University, Hisar, India

ABSTRACT

Climate change poses a major environmental challenge of the 21st century, causing long-term shifts in climate patterns that affect agriculture. This includes changes in temperature, rainfall, and carbon dioxide levels, which impact plant growth, crop yields, and food security. In India, where agriculture is vital to the economy and employs a large workforce, climate change exacerbates issues like rainfall dependence and crop profitability. To combat these challenges, India must integrate adaptation and mitigation strategies, including climate-smart agriculture and enhanced weather data. Collaboration among policymakers, researchers, and farmers is crucial to improve resilience and ensure sustainable food production and economic security.

INTRODUCTION

Agriculture plays a vital role in securing food, nutrition and livelihoods in India. While the sector has made notable advancements in the past, it currently faces numerous challenges. Key issues include stagnation in net sown area, declining soil quality, reduced per capita land availability, plateauing yield levels and negative impacts of climate change. Additionally, growing population is exerting pressure on the agricultural sector to increase food production. This task is particularly daunting as approximately 60% of the net cultivated area relies on rainfall and is vulnerable to various biotic and abiotic stresses linked to climate variability and change (Bahl, 2015). Moreover, over 80% of Indian farmers are smallholders or marginal farmers with limited capacity to adapt. The farming community in India is also diverse and largely unorganized. Climate change encompasses long-term and significant alterations in climate indicators such as rainfall, temperature, wind and snowfall patterns. Global warming and the emissions of greenhouse gases (GHGs) are identified as key contributors to the rapid intensification of climate change. The implications of climate change and its variability are likely to worsen future food security by exerting pressure on agricultural practices and their sustainability. The global climate crisis may result in the melting or reduction of glaciers, an increase in sea levels and a rise in greenhouse gas concentrations (Nema et al., 2012). Notably, global warming is the most pressing environmental challenge, primarily driven by elevated levels of atmospheric greenhouse gases, including carbon dioxide (CO_2), methane (CH_4) and nitrous oxide (N_2O) (Jain et al., 2015). These gases trap outgoing infrared radiation from the Earth's surface, consequently elevating atmospheric temperatures (Ramanathan and Feng, 2009). Recent analyses highlight a trend of increasing temperatures, a rise in the frequency of hot days and nights, more prevalent heat waves, a greater incidence of significant precipitation events, quicker snowmelt and an upward trend in sea levels. The Fifth Assessment Report from the IPCC, released in 2014 confirmed that climate system warming is indisputable. The impact of human activities on the climate is evident, as indicated by the increasing levels of greenhouse gases in the atmosphere and the resulting positive radiative forcing. Global climate change significantly affects crops, soils, livestock and pest populations (Aggarwal, 2008). This report examines the effects of climate change on agriculture in India and the technologies required to develop climate-resilient agricultural practices.

CLIMATE CHANGE IN GLOBAL AND INDIAN CONTEXT

Recent years have seen a significant increase in evidence regarding climate change based on observations. Concurrently, advancements in methods for characterizing and quantifying uncertainty have underscored the ongoing challenges in developing comprehensive long-term global and regional climate data records. The globally averaged combined land and ocean surface temperature data, derived from a linear trend, indicates a warming of 0.85 [0.65 to 1.06] °C from 1880 to 2012, based on multiple independently produced datasets and approximately 0.72°C [0.49°C to 0.89°C] from 1951 to 2012 (Stocker et al., 2013). The total increase from the average temperature of the 1850–1900 period to that of 2003–2012 is 0.78 [0.72 to 0.85] °C, while the increase from the 1850–1900 average to the reference period for projections (1986–2005) is 0.61 [0.55 to 0.67]

°C, based on the longest available dataset (IPCC, 2014). Since 1901, it is probable that precipitation levels in the mid-latitude terrestrial regions of the Northern Hemisphere have risen, exhibiting medium confidence prior to 1951 and high confidence in the years following. For other latitudinal zones, the area-averaged long-term trends exhibit low confidence due to issues with data quality, completeness, or discrepancies among available estimates. It is highly probable that the frequency of cold days and nights has diminished, while warm days and nights have become more prevalent globally since around 1950 (Alexander et al., 2006). Additionally, it is likely that heat wave occurrences have risen in significant portions of Europe, Asia and Australia during this timeframe. Furthermore, since 1950, it is probable that the frequency of heavy precipitation events on land has risen in more areas than it has fallen. It is anticipated that the global mean surface air temperature will increase by approximately 0.3 to 0.7°C, with a medium degree of certainty (Kirtman et al., 2013). It is more probable that the global mean surface air temperature during the period from 2016 to 2035 will be approximately 1°C higher than the average temperature recorded between 1850 and 1900. In terms of zonal mean precipitation, an increase is likely in high and certain mid-latitude regions, while a decrease is more probable in the subtropics (Cai et al., 2012). On a more localized scale, changes in precipitation may be affected by human-induced aerosol emissions and will be significantly influenced by natural internal variability (Risser et al., 2024). Models predict that, in the near future, there will be a rise in the duration, intensity and spatial reach of heat waves and warm spells, although the pace of these changes may differ from the overall trend of warming. It is anticipated that the average frequency and intensity of heavy rainfall events over land will increase in the near future. However, this trend may vary across different regions due to natural variability and the possible effects of human-made aerosols (Gupta and Pathak, 2016). Over the past century, India's mean annual surface air temperature has risen by 0.4 to 0.6°C (Rupa Kumar et al., 2002). Annamalai (2010) noted a downward trend in precipitation has been observed during the southwest and northeast monsoon seasons in many areas of central and northern India. In contrast, the peninsular region, especially between latitudes 9-16° N, which encompasses significant rice cultivation zones, has seen a rising trend in rainfall, particularly during the northeast monsoon season.

Sajjad (2008) projected that according to climate change forecasts for the Indian subcontinent, it is anticipated that by the 2040s, the annual mean maximum and minimum surface air temperatures will increase by 0.7°C and 1.0°C, respectively, relative to the levels observed in the 1980s. Khan et al. (2009) By 2070, a temperature rise of 0.4 to 2.0°C is expected in the *kharif* season, while the *rabi* season may see an increase of 1.1 to 4.5°C in the Indian region. The fourth assessment report from the IPCC forecasts that India will experience a temperature rise ranging from 2.7 to 4.3°C.

The IPCC Fourth Assessment Report offers persuasive evidence indicating that climate change is occurring at an accelerated pace (Munasinghe et al., 2003):

1) **Sea-level rise:** Over the last century, global sea levels have risen by around 17 cm (6.7 inches). In contrast, the rate of increase in the past decade has been nearly double that of the entire previous century.
2) **Global temperature rise:** The majority of this increase in temperature has been observed since the 1970s, with the 20 highest temperature years noted since 1981 and all 10 of the warmest years happening in the past 12 years.
3) **Warming oceans:** The oceans have taken in a significant portion of the additional heat, with the upper 700 meters (approximately 2,300 feet) of water experiencing a temperature rise of 0.302° F since 1969.
4) **Shrinking ice sheets:** The Greenland and Antarctic ice sheets have experienced a significant decrease in mass. NASA's Gravity Recovery and Climate Experiment data reveals that Greenland lost ice at a rate of 150-250 km^3 (36-60 cubic miles) per year from 2002 to 2006, while Antarctica lost approximately 152 km^3 (36 cubic miles) of ice between 2002 and 2005.
5) **Declining Arctic-sea ice:** The area and thickness of Arctic-sea ice have experienced a significant reduction in recent decades.
6) **Glacial retreat:** Glaciers are experiencing a widespread retreat globally, affecting regions such as the Andes, Alps, Alaska, Himalayas, Rockies and Africa.
7) **Ocean acidification:** The acidity of surface ocean waters has escalated by roughly 30% since the start of the Industrial Revolution. Each year,

the upper ocean layers are taking in about 2 billion tons of carbon dioxide.

CURRENT CHALLENGES:

1. **Impacts on Agriculture and Food Production:** Agricultural production in India is highly susceptible to climate variations, particularly fluctuations in monsoon rainfall and seasonal temperature changes (Lal, 2003). Research conducted by the Indian Agricultural Research Institute (IARI) and other organizations suggests that the *rabi* crop is likely to experience significant losses. For instance, increase of 1°C in temperature can lead to a reduction of 4-5 million tonnes in wheat output. Even minor alterations in temperature and rainfall can profoundly impact the quality of various crops, including fruits, vegetables, tea, coffee, aromatic and medicinal plants, as well as basmati rice (Mehraj et al., 2022). Additionally, the dynamics of pathogens and insect populations are closely linked to temperature and humidity, meaning that shifts in these factors could alter their prevalence (Skendžić et al., 2021). Various agricultural sectors might experience challenges, including possible decreases in dairy cattle yields and detrimental effects on fish breeding, migration and overall harvests. Estimates predict that global agricultural output could decline by 10% to 40% by the end of the century in 2100. The southwest monsoon significantly impacts India's climate, serving as the primary source of rainfall in the region, which is critical for both potable water and irrigation in agriculture. The productivity of agriculture is affected by two primary types of climate-related factors: (1) direct impacts resulting from alterations in temperature, carbon dioxide levels or precipitation and (2) The indirect effects arising from variations in soil moisture levels, along with the increased occurrence and intensity of pest and disease outbreaks. The vulnerability of agricultural production to climate change is influenced not only by the physiological reactions of crops but also by the ability of socio-economic systems to adjust to changes in yield and the frequency of droughts or floods.

2. **Reduction in Crop Yield:** An increase in the average temperature beyond a certain threshold will lead to a decrease in agricultural productivity (Hatfield and Prueger, 2015). The alteration in minimum temperature is more significant than that of maximum temperature (Rao et al., 2024). For instance, rice grain yield has been observed to drop by 10% for every 1° C rise in the minimum temperature during the growing season when it exceeds 32° C (Pathak et al., 2003). Research on the effects of climate change on rice productivity in Punjab, indicates that if all other climatic factors remain unchanged, a temperature rise of 1° C, 2° C and 3° C would result in reductions in rice grain yield of 5.4%, 7.4% and 25.1%, respectively (Aggarwal et al., 2009).
3. **Water Scarcity:** The rise in temperatures is expected to lead to an increase in water scarcity, with a corresponding surge in the demand for irrigation (Taheripour et al., 2015). Higher air temperatures will contribute to greater potential evapotranspiration in regions located south of 40° N (Kumar et al., 2024). Additionally, climate change is anticipated to cause a decline of approximately 20% in rice yields across India due to water shortages (Singh et al., 2024). Variability in Droughts, Cyclones, Monsoon Patterns and Floods Indian agriculture relies heavily on the timing, duration and intensity of monsoon rainfall, especially in the rainfed regions of eastern, northeastern and southern India (Pal et al., 2021). Climate models and reports from the IPCC suggest that a warmer climate may lead to greater variability in Asian Monsoon patterns. Even with an increase in irrigated land, droughts resulting from insufficient and uneven rainfall distribution remain significant climatic challenges affecting agricultural output in India (Gupta et al., 2011). The impact of droughts is expected to worsen in a warmer climate, while the frequency and intensity of flooding due to climate change will pose serious challenges for the Indian subcontinent (Sharma and Singh, 2007). Climate change will significantly impact the water cycle. The implications of climate change on agriculture are considerable. Concerns are mounting regarding soil fertility decline, increased salinity, alterations in the water table, deterioration of irrigation water quality and heightened resistance to various pesticides in northwestern India.

4. **Impact on Soil Fertility:** Climate change has significantly influenced agricultural stability, as well as ecological and ecosystem services. Soils are integral to the complex systems that contribute to climate change. The soil's response to climate change is influenced by its composition and structure. Elevated CO_2 levels enhance soil organic matter, while rising temperatures increase evapotranspiration rates, leading to diminished soil moisture. Soil-climate models indicate that higher atmospheric CO_2 levels adversely affect soil functions, including soil structure, stability, topsoil water retention, nutrient availability and erosion, found that an increase in mean annual soil temperature of up to 3.5° C in the top layer (0–10 cm) and 1.8° C on average (0–200 cm depth) results in a reduction of mean soil moisture by 0.039 kg kg^{-1} in the subsurface layer (40–100 cm) and 0.019 kg kg^{-1} on average (0–200 cm depth), corresponding to decreases of 16% and 7%, respectively. Numerous studies have shown that rising soil temperatures can disrupt soil respiration, CO_2 release, microbial processes (such as nitrification and denitrification), soil organic carbon decomposition, nitrogen use efficiency and crop yields.
5. **Impact on Nutrient Dynamics:** The integrity of soil organic matter is heavily impacted by global climate change, particularly through its effects on the carbon and nitrogen cycles. Changes in soil organic matter directly influence various soil properties, such as structure, water retention, ion exchange capacity and nutrient availability within the soil ecosystem. CO_2 fertilization can enhance plant growth, the combination of rising temperatures and elevated CO_2 levels has been associated with decreased crop yields. The effectiveness of CO_2 fertilization may not be as pronounced as previously believed, with nitrogen limitations shown to adversely impact plant growth.
6. **Impact on Pest and Predator Relation:** The 1943 Great Bengal Famine in India, triggered by a seemingly innocuous fungus, resulted in the loss of approximately three million lives (Dasgupta, 1988). Despite advancements in pest and pathogen management, contemporary farmers still face annual crop production losses of 10–16% due to biological threats (Oerke, 2006). The dynamic nature of pest-predator interactions makes it challenging to forecast their relationships amid shifting environmental conditions, as insect pests can quickly adapt

to minor environmental changes (Skendžić et al., 2021). Elevated temperatures may influence these interactions in both beneficial and detrimental ways, presenting a dual possibility.

a. **Effect of rising temperature:** In temperate climates, an increase in temperature can lead to a rise in insect populations (Robinet and Roques, 2010). Elevated temperatures could influence the survival, growth, geographic spread and population trends of insects. The physiology of insects may also be influenced by these temperature changes (Neven, 2000). Certain insects, like cicadas and Arctic moths, may take several years to complete their life cycles, while others, such as cabbage maggots and European corn borers, can develop rapidly within specific temperature thresholds (Reddy & Reddy, 2015). This can result in greater damage to crops. Additionally, migratory pests may initiate their migrations earlier due to warmer temperatures (Skendžić et al., 2021). Rising temperatures can affect the gender ratios of insects, including thrips. Warmer winters may lead to lower mortality rates among insects, contributing to population increases (Robinet and Roques, 2010). Moreover, higher temperatures may prompt crops and their pests to shift to higher altitudes.
b. **Effect of Precipitation:** The impact of rain on insect populations is significant, as it can physically remove pests like leafhoppers, plant hoppers, thrips and cutworms from their hosts (Mochida et al., 1987). Some insects, including mealybugs and the pupae of various species such as fruit flies, helicoverpa, rice stem borers, spodoptera and etiella may drown in heavy rainfall. Flooding acts as an effective management strategy for controlling both termites and stem borers. Furthermore, heavy rains can trigger pest outbreaks due to fungal pathogens, notably in sugarcane pyrilla (Choudhary et al., 2014).
c. **Effect of rising CO2 level:** Carbon dioxide serves as an example of a transformation that can yield both beneficial and detrimental outcomes. It is expected to have positive physiological impacts by enhancing photosynthesis, particularly benefiting C_3 crops like wheat and rice more than C_4 crops such as maize and grasses (Shanker et al., 2022). The fluctuations in CO_2 concentration will

manifest directly as alterations in temperature, shifts in precipitation and variations in sunlight exposure. In contrast, indirect effects will alter soil moisture levels and increase pest and disease infestations due to rising temperatures and humidity (Skendžić et al., 2021). These indirect effects, resulting from higher temperatures, are likely to shorten the duration of crops, elevate respiration rates, increase evapotranspiration, reduce fertilizer efficiency and intensify pest infestations (Malhi et al., 2021). There is a consensus that the productivity of key season (*kharif*) crops will benefit from higher levels of CO_2, while *rabi* crops are projected to experience notable declines due to the effects of rising temperatures.

7. **Impact of Climate Change on Disease:** Increases in CO_2 or climate change may lead to direct yield improvements; however, these gains could be mitigated or negated by the detrimental effects of phytophagous insects, weeds and plant pathogens (Fuhrer, 2003). It is crucial to recognize the biotic constraints on crop yields when evaluating the effects of climate change.
   a. **Impacts on Plant Patho-systems:** Climate change can significantly influence the physiology and resistance of hosts, as well as affect the developmental stages and rates of pathogens (Pathak et al., 2018). The anticipated effects include alterations in the geographical distribution of both hosts and pathogens, modifications in the physiological interactions between them and variations in crop loss.
   b. **Geographical Distribution of Host and Pathogen:** The emergence of new disease complexes is possible and certain diseases may lose their economic significance if climate warming leads to a poleward shift in agro-climatic zones, prompting host plants to migrate to new areas (Boddy, 2016). The rate at which pathogens establish themselves in a new region will depend on various factors, including the mechanisms of pathogen dispersal, the environmental suitability for such dispersal, their survival across seasons and any alterations in host physiology and ecology in the new surroundings (Gilligan & van den Bosch, 2008).
   c. **Elevated Temperature:** Rising temperatures can alter the physiology and resistance of hosts. A wealth of data exists regarding heat-

induced vulnerability and temperature-sensitive genes. Conversely, higher temperatures have been shown to promote lignification in the cell walls of forage species, thereby increasing their resistance to fungal pathogens (Zhao et al., 2014). Consequently, the effects will vary based on the specific host-pathogen interactions and the underlying resistance mechanisms. Agricultural crops and plants in natural ecosystems may serve as symptomless carriers of pathogens, with disease emergence likely if plants experience stress in a warmer climate (Singh et al., 2023). Host stress plays a critical role in the decline of numerous forest species.

8. **Impacts on Dairy, Fishery and Poultry:** Research from the National Dairy Research Institute in Karnal has revealed that without adaptation strategies, global warming could result in a decline of 1.6 million tons in milk production by 2020 and 15 million tons by 2050. Lactating cows and buffaloes experience elevated body temperatures, making it difficult for them to maintain thermal equilibrium. The body temperature of these animals can be 1.5 to 2 degrees Celsius higher than their normal levels, necessitating the implementation of more effective cooling systems to alleviate the thermal stress on lactating livestock (Khan et al., 2023). Climate change is expected to increase the water, shelter and energy needs of livestock to meet the anticipated demand for milk. Additionally, rising temperatures may significantly impact fish health and their distribution patterns. As sea surface temperatures rise, species such as the oil sardine are likely to migrate to preferred temperature zones, particularly in northern latitudes and eastern longitudes, thereby expanding their distribution and establishing fisheries in broader coastal regions (Vivekanandan et al., 2009). Historical data indicates that the Indian mackerel has adapted to rising sea temperatures by moving northward and diving to greater depths. Recently, Indian Major Carps have been observed maturing and spawning as early as March in West Bengal, with their breeding season extending from 110 to 120 days, allowing for two breeding cycles within a year at intervals of 30 to 60 days (Dey et al., 2007). In the poultry industry, heavy meat-type chickens have shown significantly higher heat stress levels compared to lighter layer and native breeds. An increase in

temperature from 31.6°C to 37.9°C has been linked to reduced feed intake and lower egg production (Wasti et al., 2020).

Figure 1. Impact of climate change on agriculture sector (Aggarwal et al., 2009)

MITIGATION AND ADAPTATION

The primary challenge in adapting and mitigating climate change within the agricultural sector lies in the necessity to produce (i) greater quantities of food, (ii) with enhanced efficiency, (iii) amidst increasingly unpredictable production conditions and (iv) while achieving net reductions in greenhouse gas emissions associated with food production and distribution (Lybbert and Sumner, 2010). Provided that climate change and the resulting policy measures do not significantly disrupt long-term income growth or drastically alter the ongoing decline in relative food prices, the overall demand for food will persistently rise in tandem with population and income increases. As climate change impacts the availability of essential inputs, particularly water in various regions, it becomes imperative to enhance input use efficiency to meet these productivity challenges. Carbon emission regulations may compel producers to acknowledge greenhouse gas emissions as a significant and costly factor in their production processes, thereby creating new opportunities and incentives for mitigating emissions on farms (Blandford and Josling, 2009). Producers will face these escalating demands and evolving incentives in the context of more unpredictable production environments. Agricultural technologies will be crucial in assisting producers to navigate these fundamental challenges. Many new technologies modify the utilization of farm inputs, often influencing how weather affects production and how production impacts carbon emissions. While the majority of agricultural technologies have climate-related implications, there are several current and emerging technologies particularly pertinent to agriculture in developing countries and its intersection with climate change.

1. **Greenhouse Gas Emission Reduction:** The continuous rise in greenhouse gas emissions, such as CO_2, CH_4 and N_2O, can be attributed to various natural phenomena and human activities, significantly impacting agricultural productivity, soil health and human well-being (Verge et al., 2007). Agricultural fields are notable contributors to these emissions, with rice paddies primarily releasing CH_4 and wheat fields emitting N_2O. Approximately 13% of global greenhouse gas emissions originate from agricultural practices, accounting for one-fifth of the annual increase in radiative forcing (IPCC, 2007). Since the preindustrial period, global concentrations of CO_2, N_2O and CH_4 have surged by 36%, 18% and 148%, respectively (IPCC 2007; Kumar et al., 2016). It is imperative to adopt sustainable agricultural methods to mitigate environmental damage. Agricultural practices also generate greenhouse gases (GHGs) and ammonia (NH_3), which can lead to acidification and eutrophication (Novak and Fiorelli, 2010). Rice farming is a significant source of both methane (CH_4) and nitrous oxide (N_2O). Increased levels of carbon dioxide can improve the rate of photosynthesis and water use efficiency in C3 plants, leading to higher crop yields. O'Mara (2012) highlighted that pasturelands have the potential to significantly contribute to the reduction of greenhouse gas emissions and enhance carbon sequestration in grazing regions.
2. **Sequestration of carbon in agricultural soils:** Reducing CO_2 emissions from agriculture can also be accomplished by enhancing carbon

sequestration in soil, which involves the accumulation of carbon as soil organic matter (Lal, 2004; Pathak et al., 2011). Effective nutrient management is essential for the sequestration of soil organic carbon (SOC) in tropical regions. Providing sufficient nutrients to the soil can boost biomass production and increase SOC levels. The application of organic manure and compost is more effective in expanding the SOC pool compared to using the same quantity of nutrients in inorganic fertilizers. Long-term application of manure can significantly raise the SOC pool, with benefits that may last for up to a century (Lal, 2013). Increased SOC not only helps in reducing CO_2 emissions but also improves soil productivity. However, it is noted that SOC sequestration poses significant challenges in tropical and sub-tropical soils, where harsh climates and limited resources prevent farmers from utilizing organic manure and crop residues. The rate of carbon mineralization is elevated in tropical regions due to high temperatures, while the efficiency of humification remains low (Guggenberger, 2005).

3. **Use of Alternate Energy Resources:** The IPCC's Special Report on Renewable Energy Sources and Climate Change Mitigation (SRREN) underscores the significance of utilizing alternative energy sources for effective climate change mitigation. In terms of energy consumption, agriculture is the third largest sector, following transportation and industry. A significant portion of energy is consumed through mechanical means (including farm machinery, human labor and animal power), chemical applications (like fertilizers, pesticides and herbicides) and electrical processes (primarily irrigation) (Choudhary and Kumar, 2013). Implementing renewable energy in agriculture could serve as an effective strategy for climate change mitigation. Although the government offers subsidies for the installation of renewable energy systems, the overall cost remains a barrier for many farmers. Agricultural practices can play a role in mitigating climate change by minimizing fossil fuel combustion, enhancing carbon storage in soil and plant biomass and reducing emissions associated with soil management. Utilizing alternative or green energy can lower energy consumption in agricultural machinery, decrease fuel requirements, shorten operational durations, reduce emissions and prolong the lifespan of equipment (Sapkota et al., 2015).

ADAPTATION STRATEGIES TO CLIMATE CHANGE

Adaptation and maladaptation:

Adaptation: Climate change adaptation involves the process of modifying practices and strategies in response to current or anticipated climatic conditions and their impacts. (UNFCCC, 2017). Adaptation in human systems is geared towards minimizing or avoiding negative impacts and leveraging positive opportunities. The adaptation strategies in agriculture are significantly intertwined with a range of other adaptation concerns across different sectors. It is essential to prevent the spread of any actions or processes that could heighten vulnerability to hazards associated with climate change, a phenomenon referred to as maladaptation. The phenomenon of maladaptation arises from development policies that prioritize certain sectors, providing short-term economic benefits while inadvertently increasing vulnerability in other sectors or over a longer time frame (Olhoff and Schaer, 2010).

1. **Climate Resilient Agriculture (CRA):** Climate resilience is a key element in the management of climate-related risks. In this framework, resilience denotes the capacity of agricultural systems to foresee, prepare for, adapt to, absorb and recover from the effects of climate change and extreme weather events. This resilience can be bolstered through the implementation of both short-term and long-term strategies for climate mitigation and adaptation, alongside fostering transparent and inclusive participation from various stakeholders in decision-making and management processes. Strengthening food security and enhancing the climate resilience of agriculture while simultaneously reducing carbon emissions or achieving carbon neutrality is essential for the Sustainable Development Goals (SDGs), particularly in relation to Goal 1 (No Poverty), Goal 2 (Zero Hunger) and Goal 3 (Climate Action). This presents a significant challenge for the agricultural sector (Pathak, 2023). Climate-resilient agriculture (CRA) represents an approach that focuses on the sustainable utilization of existing natural resources through crop and livestock production systems, aiming for long-term productivity and increased farm incomes amidst climate variability.

Conservation agriculture (CA) is being embraced by numerous countries as a leading model to enhance productivity, sustainability and climate resilience, serving as an alternative to traditional production systems. CA is founded on three interconnected principles: (1) minimizing mechanical soil disturbance and directly seeding into untilled soil to enhance soil organic matter and health, (2) increasing organic matter cover on the soil through cover crops and/or crop residues and (3) diversifying crops through associations, sequences and rotations to improve system resilience. These principles can be applied across various rainfed and irrigated production systems, including horticulture, agroforestry, organic farming, rotational farming and integrated crop-livestock systems, thereby promoting ecological sustainability.

2. **Integrated nutrient management:** Mismanagement of nutrients accounts for approximately 80% of the economic losses associated with environmental degradation stemming from both traditional and modern agricultural practices, primarily due to greenhouse gas emissions, volatilization and groundwater contamination (Subash et al., 2015). Effective nutrient management is recognized as one of the most economical strategies for promoting sustainable development, addressing climate change and enhancing the economic viability and resource efficiency of farms at both national and international levels. Integrated nutrient management (INM) involves the amalgamation of traditional and contemporary plant nutrient management techniques into a farming system that is both ecologically sustainable and economically viable, leveraging all available resources in a cohesive and efficient manner. INM improves the nitrogen use efficiency in plants while simultaneously reducing volatilization, gas emissions, leaching of inorganic pollutants and immobilization (Zhang et al., 2012; Wu and Ma 2015; Singh et al., 2017). Additionally, INM effectively manages nitrogen losses and their associated impacts while ensuring high levels of crop productivity.
3. **Soil Quality and Fertility Management:** Soil plays a crucial role in ecosystems and serves as a vital support system. Key factors contributing to soil degradation include erosion, salinization, nutrient and carbon depletion, drought and changes in soil structure. Climate change exacerbates the decline in soil quality and fertility. Rising temperatures

accelerate the decomposition of soil carbon and affect water availability. Reduced rainfall and exposed land increase the risk of wind erosion. The accelerated degradation of soil impacts the release of greenhouse gases into the atmosphere. Areas with high population density and climate change face significant threats of soil erosion (Lal, 2012). The growing population demands more food and land for housing, leading to land clearing and the expansion of agricultural areas. Anthropogenic activities driven by demographic pressures and industrial needs are among the primary causes of soil degradation. Climate model studies indicate that global warming may reduce soil moisture in semiarid regions (Sivakumar, 2007). Implementing agroforestry practices can enhance soil quality and fertility, while organic farming contributes to soil balance and boosts crop yields. The use of organic fertilizers, advanced agronomic techniques and hybrid seeds can greatly enhance both crop production and the soil ecosystem.

4. **Assessment and Evaluation of Climate-Resilient Crops:** To address the food needs of an increasing population, it is vital to greatly boost agricultural productivity. Climate change is expected to play a pivotal role in affecting crop yields. To adequately nourish this growing demographic, the development of climate-resilient crop varieties and the enhancement of agricultural systems through climate-smart practices are essential. Researchers worldwide are actively working on creating these resilient crop varieties, although this presents considerable challenges. According to Henry (2014), genomic characterization strategies can be utilized to aid in this development. A comprehensive assessment of climate change impacts, coupled with the formulation of adaptive strategies, is vital for strategic research across all agricultural sectors. The focus areas include establishing a climate-resilient agricultural system in a changing environment and advancing relevant technologies. To boost agricultural productivity, four key modules have been identified: natural resource management, soil health improvement, crop production and livestock management. India has initiated several collaborative efforts with both national and international institutions to foster a climate-resilient agricultural framework. The application of genetic engineering techniques, such as gene pyramiding, is essential for combining desirable traits into a single plant, thereby creating an

'ideal plant type' that is also resilient to adverse climatic conditions (Gaba et al., 2021). It is imperative to provide farmers with cultivars that possess a diverse genetic base, enabling them to adapt more effectively through the availability of new varieties that can withstand drought, heat and salinity, ultimately reducing the risks associated with climate variability. For instance, rice varieties such as Sahbhagi Dhan, CR Dhan 209 and CR Dhan 210 are well-suited for drought-prone areas, while Swarna sub1 excels in submerged conditions. In the case of wheat, varieties like HD 2987 demonstrate drought tolerance and DBW73 is capable of enduring high temperatures (Pathak, 2023).

5. **Water Management:** In India, the agricultural sector heavily relies on irrigation for effective crop production. Globally, approximately 60% of irrigated land utilizes surface water, while groundwater accounts for around 40% (Siebert et al., 2010). The excessive extraction of groundwater can lead to detrimental effects on aquifers and result in depletion of these vital resources. In the context of India, surface water is responsible for irrigating 24% of rice and 25% of wheat, whereas groundwater supports 49% and 72% of these crops, respectively (Smilovic et al., 2015). Drip irrigation stands out as an effective method that ensures even water distribution across fields, thereby enhancing crop yields (Benouniche et al., 2014). Additionally, pivot irrigation techniques, which simulate rainfall distribution, offer significant benefits (Kendouci et al., 2013). A shift towards drier climates is likely to reduce water availability, necessitating training for farmers in modern irrigation practices. Furthermore, rainwater harvesting programs are essential for replenishing groundwater supplies and enhancing irrigation practices. Olayide et al., (2016) emphasized the importance of minimizing climate-induced production risks through the adoption of climate-smart agriculture (CSA). This approach not only improves irrigation but also promotes sustainable agricultural water management, contributing to food security. Enhancing the resilience of agriculture to climate change through effective irrigation management can lead to sustainable agricultural production (Nwanze and Fan 2016; Olayide et al., 2016).

Water harvesting, a time-honored technique in India, presents a viable approach to address the challenges of irrigation water scarcity throughout the year. This method has demonstrated considerable water conservation benefits (Pandey et al., 2003). In addition, the implementation of micro-irrigation systems, including sprinklers and drip irrigation (Raihan, 2024), can lead to substantial reductions in water usage. Furthermore, laser land leveling (LLL) enhances the efficiency of water and nutrient utilization, thereby bolstering agricultural resilience against water stress. Laser leveling creates a smooth, even field that optimizes water distribution and minimizes losses. This method also ensures uniform placement of seeds and fertilizers, promoting better plant establishment, improved nutrient efficiency and higher yields. The practice of zero tillage in wheat cultivation offers both economic advantages and environmental gains (Somasundaram et al., 2020). Furthermore, in agroforestry systems, the breakdown of leaf litter plays a significant role in enhancing soil fertility, reducing water runoff and preventing erosion, which collectively strengthens resilience against climate variability.

6. **Disease and Pest Control:** It has been estimated that approximately 40% of global crop losses are attributed to pests (Oerke, 2006). The distribution of pests, including insects and fungal and bacterial pathogens, is influenced by global climate change. Pests have the ability to adjust their body temperature across a broad range of environmental conditions. The physiology of insects is primarily influenced by temperature variations. Increased temperatures or higher levels of CO_2 can accelerate the life cycles of pathogenic fungi. Microclimatic variations are particularly important for pests due to their small size. Additionally, factors such as plant height and canopy structure, which are influenced by microclimate, can create barriers to the dispersal of pathogens (Ando and Grumet, 2006). Furthermore, the metabolism and movement of pesticides are impacted by the morphological and physiological changes occurring within the plant canopy (Pangga et al., 2013). Sustainable agricultural production can be supported through the management of pests and diseases by using resistant varieties, natural pesticides, bacterial and viral agents and pheromones to disrupt pest reproduction. The role of bioagents in pest management is

vital and practices that enhance natural predators and parasites, such as creating habitats for them, establishing beetle banks and flowering strips and implementing crop rotation and intercropping, should be incorporated into pest management strategies. Additionally, reducing pesticide usage can contribute to lower carbon emissions.

7. **Crop Diversification:** Advocating for the diversification of crop and livestock varieties, including the substitution of existing plant types, cultivars, hybrids and animal breeds with those that offer improved drought and heat tolerance, is seen as a viable strategy to enhance productivity in response to temperature and moisture challenges (Reddy and Reddy, 2015). The genetic diversity within seeds has been acknowledged as a powerful defense mechanism against diseases, pests and climate-related risks. Furthermore, the increasing demand for high-value food items—such as fruits, vegetables, dairy products, meat, eggs and fish—due to rising incomes and urbanization is leading to a decline in the consumption of traditional staples like rice and wheat (Pingali et al., 2019). Shifting from rice-wheat systems to high-value crops can boost farmers' income and decrease the use of water and fertilizers. One effective approach to bolster the resilience of agricultural systems against climate change is through the diversification of crops, both in variety and planting schedules (Rao et al., 2014). Increased diversity within production systems enhances their resilience, thereby supporting food security and improving nutritional outcomes (Jat et al., 2007; Gora et al., 2022). Furthermore, diverse agricultural practices are vital for delivering essential ecosystem services, including nutrient cycling, carbon sequestration, erosion prevention, greenhouse gas mitigation and the regulation of water cycles. By diversifying crops, farmers can better withstand climate change impacts, as this strategy helps control pest populations and minimizes the risk of disease spread associated with climate variability, ultimately protecting crop yields from climatic pressures.
8. **Changes in Land-use Management Practices:** Modifying land-use strategies, including the placement of crop and livestock operations, rotating production between different crops and livestock and moving production away from less productive areas, can significantly mitigate the risks posed by climate change to agricultural output. Additionally,

adjusting the cropping sequence—such as altering the timing of sowing, planting, spraying and harvesting—can optimize the benefits of shifting growing seasons and varying moisture and heat conditions (Pathak et al., 2012). Furthermore, changing the timing of field sowing or planting allows farmers to better align the growing season with the new environmental conditions. Adaptation strategies may also encompass adjusting irrigation schedules and the application of inputs like fertilizers.

9. **Efficient Use of Resources:** Resource-conserving technologies (RCTs) refer to practices that improve the efficiency of resource and input usage while delivering immediate and measurable economic advantages. These benefits include reduced production costs, decreased requirements for water, fuel and labor and timely crop establishment, all of which contribute to enhanced yields (Ramesh et al., 2016). In environments where heat and water stress are prevalent, the adoption of RCTs can significantly increase wheat yields, particularly on small and medium-sized farms, while also minimizing negative environmental impacts. Techniques such as zero-tillage (ZT) enable farmers to plant wheat shortly after harvesting rice, allowing the crop to mature before the onset of the hot premonsoon season. As regional temperatures continue to rise, the importance of early sowing for wheat will only increase. Field studies indicate that farmers in the rice-wheat belt of the Indo-Gangetic Plains are increasingly embracing RCTs due to their labor-saving, water-saving and early planting benefits. Additionally, RCTs in the rice-wheat system play a crucial role in reducing greenhouse gas emissions and adapting to climate change (Pathak et al., 2011). To achieve broader adaptation to climate change, these crop management strategies should be integrated with crop improvement measures.
10. **Better Weather Forecasting and Crop Insurance Schemes:** Weather forecasting and early warning systems play a crucial role in mitigating the risks associated with climate-related challenges. The integration of information and communication technologies (ICT) can significantly assist researchers and administrators in formulating effective contingency plans. It is essential to develop robust crop insurance programs that support farmers in minimizing the financial impact of crop failures

caused by adverse weather events (Pathak et al., 2012). A combination of formal, informal, private and public insurance initiatives should be established to alleviate income losses stemming from climate change. However, comprehensive information is necessary to create policies that promote effective insurance solutions. Micro-finance has proven beneficial for rural populations, particularly women and providing low-cost financial services can greatly assist vulnerable farmers. The expanding mobile telephony network can enhance SMS-based banking services, facilitating better connections between farmers and financial institutions (Chhachhar et al., 2014). Nevertheless, the availability of micro-insurance options remains limited compared to micro-finance. Therefore, it is imperative to create a sustainable insurance framework while educating rural communities about the benefits of these opportunities.

Figure 2. Climate change adaptation technologies (Prasad et al., 2019)

ENVIRONMENTAL POLICIES FOR CLIMATE CHANGE IN INDIA

India is a leading global producer of agricultural goods, with over 50% of its population employed in the agricultural sector. However, the sector is facing significant risks due to climate change, including unpredictable rainfall, rising temperatures and severe weather phenomena. To address these threats, the Indian government has crafted and put into action several policies and initiatives focused on strengthening the agricultural sector's resilience to climate change and its associated impacts. Each of these initiatives has specific targets aimed at addressing the challenges that climate change presents to agriculture.

National Action Plan on Climate Change (NAPCC)

The NAPCC represents an extensive strategy that specifies approaches for addressing climate change through adaptation and mitigation efforts. It consists of eight primary missions. (1) the Jawaharlal Nehru National Solar Mission (JNNSM), (2) National Water Mission,, (3) National Mission on Sustainable Habitat, (4) National Mission for "Green India," (5) National Mission on Strategic Knowledge for Climate Change (NMSKCC), (6) National Mission for Enhanced Energy Efficiency (NMEEE), (7) National Mission for Sustainable Agriculture (NMSA) and (8) National Mission for Sustainable Himalayan Ecosystem (NMSHE). To address the impacts of climate change, the Indian government has established and executed a series of policies and initiatives focused on strengthening the agricultural sector. Each of these measures has specific aims to counteract the effects of climate change on agricultural practices.

1. **National Action Plan on Climate Change (NAPCC):** The National Action Plan on Climate Change (NAPCC) outlines India's strategy for addressing climate change across various sectors, including agriculture. It emphasizes the importance of developing adaptation strategies, recognizing that agriculture is especially susceptible to the impacts of climate change. The plan includes specific missions aimed at promoting climate-resilient agricultural practices and effective water

management, such as 'National Water Mission' and 'National Mission for Sustainable Agriculture (NMSA) (Government of India, 2008).

2. **Pradhan Mantri Krishi Sinchayee Yojana (PMKSY):** PMKSY employs a range of strategies such as micro-irrigation, rainwater harvesting and watershed management to increase water use efficiency. This program facilitates agricultural adaptation to climate change by adopting effective water management techniques, particularly in regions that are struggling with limited water resources (PKVY, Ministry of Jal Shakti, 2015).
3. **Paramparagat Krishi Vikas Yojana (PKVY):** PKVY endorses organic farming practices due to their proven benefits in improving soil health and promoting biodiversity. Additionally, organic farming plays a significant role in addressing climate change by minimizing the reliance on chemical inputs such as pesticides and fertilizers. Under the PKVY initiative, farmers transitioning to organic methods receive financial support and training (Ministry of Agriculture and Farmers Welfare, 2016).
4. **Soil Health Card Scheme:** Launched in 2015, this initiative focuses on assessing the health of soil and offering personalized guidance to farmers on improving soil fertility. The program is vital for adaptation, as healthy soils are better equipped to withstand the impacts of climate change, including droughts and floods (Ministry of Agriculture and Farmers Welfare, 2015).
5. **National Mission on Sustainable Agriculture (NMSA):** NMSA focuses on advancing the quality and production of food crops in order to meet the challenges posed by shifting environmental factors. Researchers are actively seeking new crop varieties that are resilient to climate change, such as those that can withstand higher temperatures, while also working to improve alternative cropping systems within the NMSA framework. The primary objective of this mission is to advance agricultural practices through various adaptation and mitigation strategies. To support this aim, several initiatives have been launched, including (1) Rainfed Area Development (RAD), (2) On-Farm Water Management (OFWM) and (3) Soil Health Management, among others (Rockström et al., 2010). The NMSA firmly supports the implementation of crop diversification, integrated farming systems

and agroforestry as viable strategies for promoting climate-resilient agriculture. By adopting climate-smart technologies and diversifying their crop selections, farmers can reduce their vulnerability to the impacts of climate change (Ministry of Agriculture and Farmers Welfare, 2014).

6. **Crop Insurance Program based on Weather (WBCIS):** Farmers are provided with financial protection through WBCIS to mitigate the impact of crop losses caused by adverse weather conditions. This initiative aims to mitigate the financial risks associated with climate variability, thereby promoting the adoption of strategies by farmers that minimize risk and strengthen climate resilience (Ministry of Agriculture and Farmers Welfare, 2016).
7. **Pradhan Mantri Fasal Bima Yojana (PMFBY):** Initiated in 2016, the Pradhan Mantri Fasal Bima Yojana (PMFBY) focuses on offering budget-friendly crop insurance to support farmers. This initiative not only supports recovery from agricultural losses but also promotes the adoption of climate-resilient practices, as the benefits of insurance are tied to the implementation of these suggested strategies (Ministry of Agriculture and Farmers Welfare, 2016).

Consequences And Difficulties In Agricultural Policy

Despite a reduction in its contribution to the national output, agriculture still stands as the primary sector of the Indian economy. Its share has fallen from over 50% in the years immediately following Independence in 1947 to under 25% in contemporary times. Approximately 58% of the workforce still relies on agriculture as their main source of employment. The sector faces significant challenges, primarily due to inadequate infrastructure and limited funding. Additionally, India's border measures, coupled with output price support, have resulted in elevated domestic food prices. To alleviate the burden of these rising prices on consumers, food subsidies have been implemented. However, input subsidies have led to overuse of agricultural inputs, causing various environmental issues, including soil salinity and depletion of groundwater resources. The escalating costs of input and food subsidies have significantly contributed to the fiscal deficits experienced by numerous states. Allocating funds for subsidies could have been more advantageous if directed towards research, education and infrastructure development, which would significantly boost productivity and competitiveness in the industry. Moreover, India's frameworks related to institutions, policies and regulations have primarily promoted the cultivation of staple foods, thereby obstructing vertical integration in the value chain and limiting the expansion of the horticultural industry. Existing policies have generally favoured specific interest groups that resist effective reform. In the face of climate change, one of the major challenges confronting both the government and the worldwide scientific community is satisfying the food demands of an increasing population. Agriculture plays a crucial role in the Indian economy and its success is largely contingent upon the monsoon rainfall. With around 58% of the workforce employed in agriculture, the adverse effects of climate change on the economy are profound. While the Government of India has initiated numerous programs to address climate change, there is an urgent need to prioritize the establishment of a "climate-smart agriculture system" through effective policies for mitigation and adaptation. Moreover, additional research is required to support innovations aimed at enhancing sustainable and climate-resilient agricultural productivity.

Strategies for Farmers to Address Climate Change Awareness

Farmers need to be aware of climate variability and change, including how these factors impact crop production and the various adaptation strategies that can be employed:

1. **Agromet advisories:** Serve the agricultural community by providing tailored bulletins that consider the current weather, soil conditions and crop status, along with forecasts. These advisories suggest measures to reduce losses and enhance input efficiency, covering aspects such as land preparation, crop and cultivar selection, sowing and harvesting timelines, irrigation planning and the application of pesticides and fertilizers, especially in light of extreme weather events. Each advisory bulletin is structured into three main components:

a. A review of weather occurrences from the previous week and a five-day forecast, which includes details on cloud cover, rainfall, wind direction, average wind speed, relative humidity and temperature ranges;
b. A report on the current growth stage of crops, ongoing farming practices and any pest or disease outbreaks; and
c. Supplementary information on recommended farm activities based on the weather forecast. The 23 State Agromet Service Centers work in conjunction with the State Department of Agriculture to issue these advisories on Tuesdays and Fridays, which are then shared through All India Radio (AIR), SMS, Doordarshan, websites and print media.

2. **Weather Index-Based Insurance Product:** The fundamental concept of weather insurance involves assessing the percentage change in crop yield resulting from unfavorable weather conditions. This type of insurance serves as a safeguard against increased costs or lost profits stemming from specific adverse weather occurrences.
3. **Contingency Planning:** This refers to a strategy developed to address exceptional risks that are difficult or impossible to circumvent. The Ministry of Agriculture, in partnership with the Indian Council of Agricultural Research (ICAR) and State Agricultural Universities (SAUs), is developing contingency plans tailored to individual districts for agriculture and associated sectors, such as fisheries and animal husbandry. This initiative began in March 2010 as part of the Rashtriya Krishi Vikas Yojana (RKVY). Plans have been created and validated for approximately 300 of the over 600 districts and the information is available on the websites of the Department of Agriculture and Cooperation (DAC), the Indian Council of Agricultural Research (ICAR) and the Centre for Research in Dryland Areas (CRIDA). Each comprehensive district-specific document outlines the crops and farming practices to implement in response to scenarios such as insufficient or delayed monsoon rains, unexpected rainfall, frost, extreme temperatures and excessive rainfall. These documents serve as scientific resources for adaptation in the face of such challenges.

CONCLUSIONS

Climate change and its increasing variability are expected to exacerbate future food security challenges by placing additional stress on agricultural systems. However, significant uncertainties persist regarding the evaluation of climate change's impacts, as well as the adaptation and mitigation strategies in agriculture. This uncertainty largely stems from the lack of standardized methodologies, which can often lead to inaccuracies and imprecision in assessments. The conversation surrounding climate change adaptation and mitigation in Indian agriculture highlights the pressing urgency and intricate nature of addressing climate-related issues within this sector. Climate change significantly threatens food security, agricultural productivity and the achievement of the Sustainable Development Goals (SDGs). These consequences are far-reaching, impacting a range of areas from inconsistent weather patterns that challenge conventional farming techniques to difficulties such as limited water resources and the degradation of soil quality. The link between climate change and food security highlights the urgent need for measures to protect vulnerable populations and ensure access to nutritious food. The agricultural sector can enhance its resilience to climate disruptions by adopting strategies like crop diversification, water-efficient methods and resilient crop varieties, while also pursuing mitigation efforts to reduce greenhouse gas emissions through organic farming, agroforestry and better livestock management. Proper implementation of policies is key to facilitating climate change strategies in India's agricultural sector.

REFERENCES

Aggarwal, P. K. (2008). Global climate change and Indian agriculture: Impacts, adaptation and mitigation. *Indian Journal of Agricultural Sciences*, *78*(11), 911.

Aggarwal, P. K., Joshi, H. C., Singh, S. D., Bhatia, A., Jain, N., Prasad, S., Chaudhary, A., Gupta, N., & Pathak, H. (2009). Agriculture and environment. In *Hand Book of Agriculture, Directorate of Information and Publication Agriculture* (pp. 62–92). ICAR.

Aggarwal, P. K., Singh, A. K., Samra, J. S., Singh, G., Gogoi, A. K., Rao, G. G. S. N., & Ramakrishna, Y. S. (2009). Introduction. In Aggarwal, P. K. (Ed.), *Global Climate Change and Indian Agriculture* (pp. 1–5). ICAR.

Alexander, L. V., Zhang, X., Peterson, T. C., Caesar, J., Gleason, B., Klein Tank, A. M. G., Haylock, M., Collins, D., Trewin, B., Rahimzadeh, F., Tagipour, A., Rupa Kumar, K., Revadekar, J., Griffiths, G., Vincent, L., Stephenson, D. B., Burn, J., Aguilar, E., Brunet, M., & Vazquez-Aguirre, J. L. (2006). Global observed changes in daily climate extremes of temperature and precipitation. *Journal of Geophysical Research*, *111*(D5), 2005JD006290. DOI: 10.1029/2005JD006290

Ando, K., & Grumet, R. (2006). Evaluation of altered cucumber plant architecture as a means to reduce *Phytophthora capsici* disease incidence on cucumber fruit. *Journal of the American Society for Horticultural Science*, *131*(4), 491–498. DOI: 10.21273/JASHS.131.4.491

Annamalai, H. (2010). Moist dynamical linkage between the equatorial Indian Ocean and the South Asian monsoon trough. *Journal of the Atmospheric Sciences*, *67*(3), 589–610. DOI: 10.1175/2009JAS2991.1

Bahl, P. N. (2015). Climate change and pulses: Approaches to combat its impact. *Agricultural Research*, *4*(2), 103–108. DOI: 10.1007/s40003-015-0163-9

Benouniche, M., Kuper, M., Hammani, A., & Boesveld, H. (2014). Making the user visible: Analysing irrigation practices and farmers' logic to explain actual drip irrigation performance. *Irrigation Science*, *32*(6), 405–420. DOI: 10.1007/s00271-014-0438-0

Blandford, D. and Josling, T. (2009). Greenhouse gas reduction policies and agriculture: implications for production incentives and international trade disciplines.

Boddy, L. (2016). Fungi, ecosystems and global change. In *The fungi* (pp. 361–400). Academic Press. DOI: 10.1016/B978-0-12-382034-1.00011-6

Cai, W., Cowan, T., & Thatcher, M. (2012). Rainfall reductions over Southern Hemisphere semi-arid regions: The role of subtropical dry zone expansion. *Scientific Reports*, *2*(1), 702. DOI: 10.1038/srep00702

Chhachhar, A. R., Qureshi, B., Khushk, G. M., & Maher, Z. A. (2014). Use of mobile phone among farmers for agriculture information. *European Journal of Scientific Research*, *119*(2), 265–271.

Choudhary, A. K., Amrate, P. K., & Chatterjee, A. (2014). An unusual incidence and population dynamics of sugarcane leaf hopper, *Pyrilla perpusilla* (Walk) and its natural enemies in Madhya Pradesh, India. *Environmental Conservation*, *20*, 337–342.

Choudhary, V. K., & Kumar, P. S. (2013). Crop and water productivity, profitability and energy consumption pattern of a maize-based crop sequence in the North Eastern Himalayan Region, India. *Archives of Agronomy and Soil Science*, *59*(5), 653–669. DOI: 10.1080/03650340.2012.665992

Dasgupta, M. K. (1988). *Principles of plant pathology*. Allied Publishers.

Dey, S., Srivastava, P. K., Maji, S., Das, M. K., Mukhopadhyaya, M. K., & Saha, P. K. (2007). Impact of climate change on the breeding of Indian major carps in West Bengal. *Journal of the Inland Fisheries Society of India*, *39*(1), 26–34.

Fuhrer, J. (2003). Agroecosystem responses to combinations of elevated CO_2, ozone and global climate change. *Agriculture, Ecosystems & Environment*, *97*(1-3), 1–20. DOI: 10.1016/S0167-8809(03)00125-7

Gaba, Y., Pareek, A., & Singla-Pareek, S. L. (2021). Raising climate-resilient crops: Journey from the conventional breeding to new breeding approaches. *Current Genomics*, *22*(6), 450–467. DOI: 10.2174/1389202922666210928151247

Gilligan, C. A., & van den Bosch, F. (2008). Epidemiological models for invasion and persistence of pathogens. *Annual Review of Phytopathology*, *46*(1), 385–418. DOI: 10.1146/annurev.phyto.45.062806.094357

Gora, M. K., Kumar, S., Jat, H. S., Kakraliya, S. K., Choudhary, M., Dhaka, A. K., Jat, R. D., Kakraliya, M., Sharma, P. C., & Jat, M. L. (2022). Scalable diversification options delivers sustainable and nutritious food in Indo-Gangetic plains. *Scientific Reports*, *12*(1), 14371. DOI: 10.1038/s41598-022-18156-1

Guggenberger, G. (2005). Humification and mineralization in soils. In *Microorganisms in soils: roles in genesis and functions* (pp. 85–106). Springer Berlin Heidelberg. DOI: 10.1007/3-540-26609-7_4

Gupta, A., & Pathak, H. (2016). *Climate change and agriculture in India.*

Gupta, A. K., Tyagi, P., & Sehgal, V. K. (2011). Drought disaster challenges and mitigation in India: Strategic appraisal. *Current Science*, •••, 1795–1806.

Hatfield, J. L., & Prueger, J. H. (2015). Temperature extremes: Effect on plant growth and development. *Weather and Climate Extremes*, *10*, 4–10. DOI: 10.1016/j.wace.2015.08.001

Henry, R. J. (2014). Genomics strategies for germplasm characterization and the development of climate resilient crops. *Frontiers in Plant Science*, *5*, 68. DOI: 10.3389/fpls.2014.00068

IPCC. (2007). *Climate Change 2007: Synthesis Report. Contribution of Working Group I, II and III to the Fourth Assessment Report of the Intergovernmental Panel on Climate Change. Fourth Assessment Report.* Cambridge University Press.

IPCC. (2014). Climate change 2014 synthesis report. *IPCC: Geneva, Szwitzerland*, 1059-1072.

Jain, N., Bhatia, A., Pathak, H., Gupta, N., Sharma, D. K. and Kaushik, R. (2015). Greenhouse gas emission and global warming. *Introduction to environmental sciences*, 379-411.

Jat, M. L., Sharma, S. K., Cothari, A. K., Jain, L. K., & Balyan, J. K. (2007). Management of drought through crop diversification and intercropping based on rainfall analysis in Western India. *Indian Journal Dryland Agricultural Research Development, 22*(2), 153–162.

Kendouci, M. A., Bendida, A., Khelfaoui, R., & Kharroubi, B. (2013). The impact of traditional irrigation (Foggara) and modern (drip, pivot) on the resource non-renewable groundwater in the Algerian Sahara. *Energy Procedia, 36*, 154–162. DOI: 10.1016/j.egypro.2013.07.018

Khan, I., Mesalam, A., Heo, Y. S., Lee, S. H., Nabi, G., & Kong, I. K. (2023). Heat stress as a barrier to successful reproduction and potential alleviation strategies in cattle. *Animals (Basel), 13*(14), 2359. DOI: 10.3390/ani13142359

Khan, S. A., Kumar, S., Hussain, M. Z., & Kalra, N. (2009). Climate change, climate variability and Indian agriculture: impacts vulnerability and adaptation strategies. In *Climate change and crops* (pp. 19–38). Springer Berlin Heidelberg. DOI: 10.1007/978-3-540-88246-6_2

Kirtman, B., Power, S. B., Adedoyin, A. J., Boer, G. J., Bojariu, R., Camilloni, I., Doblas-Reyes, F., Fiore, A. M., Kimoto, M., Meehl, G. and Prather, M. (2013). Near-term climate change: projections and predictability.

Kumar, P., Deka, D., Yadav, A., Kumar, M., Das, J. P., Singh, A., & Gurjar, A. (2024). Potential and actual evapotranspiration and Landsat derived indices. *Global Journal of Environmental Science and Management, 10*(3), 1227–1248.

Kumar, V., Kumar, S. S., & Kumar, S. (2016). A study on methane emissions and its mitigation strategies in present scenario. *SAMRIDDHI: A Journal of Physical Sciences. Engineering and Technology, 8*(02), 87–92.

Lal, M. (2003). Global climate change: India's monsoon and its variability. *Journal of Environmental Studies and Policy, 6*(1), 1–34.

Lal, R. (2004). Soil carbon sequestration in India. *Climatic Change*, *65*(3), 277–296. DOI: 10.1023/B:CLIM.0000038202.46720.37

Lal, R. (2012). Climate change and soil degradation mitigation by sustainable management of soils and other natural resources. *Agricultural Research*, *1*(3), 199–212. DOI: 10.1007/s40003-012-0031-9

Lal, R. (2013). Intensive agriculture and the soil carbon pool. *Journal of Crop Improvement*, *27*(6), 735–751. DOI: 10.1080/15427528.2013.845053

Lybbert, T. and Sumner, D. (2010). Agricultural technologies for climate change mitigation and adaptation in developing countries: policy options for innovation and technology diffusion.

Malhi, G. S., Kaur, M., & Kaushik, P. (2021). Impact of climate change on agriculture and its mitigation strategies: A review. *Sustainability (Basel)*, *13*(3), 1318. DOI: 10.3390/su13031318

Mehraj, S., Manzoor, M. M. M., Sharma, R. K., Mir, A. H., Khan, I. L., & Maqbool, S. (2022). Climate change and agriculture. In *Environmental Studies and Climate Change* (pp. 231–239). CRC Press. DOI: 10.1201/9781003220824-17

Ministry of Agriculture and Farmers Welfare. (2014). *National Mission on Sustainable Agriculture*. NMSA.

Ministry of Agriculture and Farmers Welfare. (2015). Soil Health Card Scheme. https://soilhealth.dac.gov.in/home

Ministry of Agriculture and Farmers Welfare. (2016). Paramparagat Krishi Vikas Yojana (PKVY). https://nmsa.dac.gov.in/ pdfDoc/SHM_Guidelines472016.pdf

Ministry of Agriculture and Farmers Welfare. (2016). Pradhan Mantri Fasal Bima Yojana (PMFBY). https://www.pmfby.gov.in

Ministry of Agriculture and Farmers Welfare. (2016). *Weather based Crop Insurance Scheme*. WBCIS.

Ministry of Jal Shakti. (2015). Pradhan Mantri Krishi Sinchayee Yojana (PMKSY). https://pmksy-mowr.nic.in

Mochida, O. M., Joshi, R. C. and Litsinger, J. A. (1987). Climatic factors affecting the occurrence of insect pests. *Weather and rice*, 149-164.

Munasinghe, M., Canziani, O., Davidson, O., Metz, B., Parry, M., & Harisson, M. (2003). *Integrating sustainable development and climate change in the IPCC Fourth Assessment Report*. Munasinghe Institute for Development.

Nema, P., Nema, S., & Roy, P. (2012). An overview of global climate changing in current scenario and mitigation action. *Renewable & Sustainable Energy Reviews*, *16*(4), 2329–2336. DOI: 10.1016/j.rser.2012.01.044

Neven, L. G. (2000). Physiological responses of insects to heat. *Postharvest Biology and Technology*, *21*(1), 103–111. DOI: 10.1016/S0925-5214(00)00169-1

Novak, S. M., & Fiorelli, J. L. (2010). Greenhouse gases and ammonia emissions from organic mixed crop-dairy systems: A critical review of mitigation options. *Agronomy for Sustainable Development*, *30*(2), 215–236. DOI: 10.1051/agro/2009031

Nwanze, K. F., & Fan, S. (2016). Climate change and agriculture: strengthening the role of smallholders. In *Global food policy report* (pp. 13–21). International Food Policy Research Institute.

O'Mara, F. P. (2012). The role of grasslands in food security and climate change. *Annals of Botany*, *110*(6), 1263–1270. DOI: 10.1093/aob/mcs209

Oerke, E. C. (2006). Crop losses to pests. *Journal of Agricultural Science*, *144*(1), 31–43. DOI: 10.1017/S0021859605005708

Olayide, O. E., Tetteh, I. K., & Popoola, L. (2016). Differential impacts of rainfall and irrigation on agricultural production in Nigeria: Any lessons for climate-smart agriculture. *Agricultural Water Management*, *178*, 30–36. DOI: 10.1016/j.agwat.2016.08.034

Olhoff, A., & Schaer, C. (2010). *Screening Tools and Guidelines to Support the Mainstreaming of Climate Change Adaptation into Development Assistance – A Stocktaking Report*. UNDP.

Pal, L., Ojha, C. S. P., & Dimri, A. P. (2021). Characterizing rainfall occurrence in India: Natural variability and recent trends. *Journal of Hydrology (Amsterdam), 603*, 126979. DOI: 10.1016/j.jhydrol.2021.126979

Pandey, D. N., Gupta, A. K., & Anderson, D. M. (2003). Rainwater harvesting as an adaptation to climate change. *Current Science*, •••, 46–59.

Pangga, I. B., Hanan, J., & Chakraborty, S. (2013). Climate change impacts on plant canopy architecture: Implications for pest and pathogen management. *European Journal of Plant Pathology, 135*(3), 595–610. DOI: 10.1007/s10658-012-0118-y

Pathak, H. (2023). Impact, adaptation, and mitigation of climate change in Indian agriculture. *Environmental Monitoring and Assessment, 195*(1), 52. DOI: 10.1007/s10661-022-10537-3

Pathak, H., Aggarwal, P. K., & Singh, S. D. (2012). Climate change impact, adaptation and mitigation in agriculture: methodology for assessment and applications. *Indian Agricultural Research Institute, New Delhi*, 302, 19.

Pathak, H., Byjesh, K., Chakrabarti, B., & Aggarwal, P. K. (2011). Potential and cost of carbon sequestration in Indian agriculture: Estimates from long-term field experiments. *Field Crops Research, 120*(1), 102–111. DOI: 10.1016/j.fcr.2010.09.006

Pathak, H., Ladha, J. K., Aggarwal, P. K., Peng, S., Das, S., Singh, Y., Singh, B., Kamra, S. K., Mishra, B., Sastri, A. S. R. A. S., Aggarwal, H. P., Das, D. K., & Gupta, R. K. (2003). Trends of climatic potential and on-farm yields of rice and wheat in the Indo-Gangetic Plains. *Field Crops Research, 80*(3), 223–234. DOI: 10.1016/S0378-4290(02)00194-6

Pathak, R., Singh, S. K., Tak, A., & Gehlot, P. (2018). Impact of climate change on host, pathogen and plant disease adaptation regime: A review. *Biosciences Biotechnology Research Asia, 15*(3), 529–540. DOI: 10.13005/bbra/2658

Pingali, P., Aiyar, A., Abraham, M., Rahman, A., Pingali, P., Aiyar, A., Abraham, M., & Rahman, A. (2019). Diet diversity and the declining importance of staple grains. *Transforming food systems for a rising India*, pp.73-91.

Prasad, R. S., Mohapatra, T., & Rao, S. (2019). *Climate change and Indian agriculture: impacts, coping strategies, programmes and policy*. Indian Council of Agricultural Research.

Raihan, A. (2024). A review of recent advances, challenges and potential future directions of climate-smart agriculture. In *Proceedings of The International Conference on Climate-Smart Agriculture*.

Ramanathan, V., & Feng, Y. (2009). Air pollution, greenhouse gases and climate change: Global and regional perspectives. *Atmospheric Environment*, *43*(1), 37–50. DOI: 10.1016/j.atmosenv.2008.09.063

Ramesh, R., Negi, S. C., & Rana, S. S. (2016). Resource conservation technologies (RCTs)-needs and future prospects. *RE:view*.

Rao, B. B., Chowdary, P. S., Sandeep, V. M., Rao, V. U. M., & Venkateswarlu, B. (2014). Rising minimum temperature trends over India in recent decades: Implications for agricultural production. *Global and Planetary Change*, *117*, 1–8.

Rao, C. S., Baral, K., Chanadana, V. M., Jagadesh, M., & Karthik, R.V. MANI CHANADANAM. JAGADESHR. KARTHIK. (2024). Climate change adaptation and mitigation in Indian agriculture. *Journal of Agrometeorology*, *26*(2), 137–148. DOI: 10.54386/jam.v26i2.2582

Reddy, P. P. and Reddy, P. P. (2015). Climate change adaptation. *Climate resilient agriculture for ensuring food security*, 223-272.

Risser, M. D., Collins, W. D., Wehner, M. F., O'Brien, T. A., Huang, H., & Ullrich, P. A. (2024). Anthropogenic aerosols mask increases in US rainfall by greenhouse gases. *Nature Communications*, *15*(1), 1318. DOI: 10.1038/s41467-024-45504-8

Robinet, C., & Roques, A. (2010). Direct impacts of recent climate warming on insect populations. *Integrative Zoology*, *5*(2), 132–142. DOI: 10.1111/j.1749-4877.2010.00196.x

Rockström, J., Karlberg, L., Wani, S. P., Barron, J., Hatibu, N., Oweis, T., Bruggeman, A., Farahani, J., & Qiang, Z. (2010). Managing water in rainfed agriculture—The need for a paradigm shift. *Agricultural Water Management*, *97*(4), 543–550. DOI: 10.1016/j.agwat.2009.09.009

Rupa Kumar, K., Ashrit, R. G., & Pant, G. B. (2002). The Indian summer monsoon: Past, present and future. *Science and Culture*, *68*(9-12), 217–224.

Sajjad, S. H. 2008. Impacts of climate change on south asia.

Sapkota, T. B., Jat, M. L., Aryal, J. P., Jat, R. K., & Khatri-Chhetri, A. (2015). Climate change adaptation, greenhouse gas mitigation and economic profitability of conservation agriculture: Some examples from cereal systems of Indo-Gangetic Plains. *Journal of Integrative Agriculture*, *14*(8), 1524–1533. DOI: 10.1016/S2095-3119(15)61093-0

Shanker, A. K., Gunnapaneni, D., Bhanu, D., Vanaja, M., Lakshmi, N. J., Yadav, S. K., Prabhakar, M., & Singh, V. K. (2022). Elevated CO_2 and water stress in combination in plants: Brothers in arms or partners in crime. *Biology (Basel)*, *11*(9), 1330. DOI: 10.3390/biology11091330

Sharma, K. D., & Singh, P. (2007). Impact of climate change on hydrological extremes: Floods and droughts. *Hydrology Journal*, *30*(34), 129–145.

Siebert, S., Burke, J., Faures, J. M., Frenken, K., Hoogeveen, J., Döll, P., & Portmann, F. T. (2010). Groundwater use for irrigation–a global inventory. *Hydrology and Earth System Sciences*, *14*(10), 1863–1880. DOI: 10.5194/hess-14-1863-2010

Singh, B. K., Delgado-Baquerizo, M., Egidi, E., Guirado, E., Leach, J. E., Liu, H., & Trivedi, P. (2023). Climate change impacts on plant pathogens, food security and paths forward. *Nature Reviews. Microbiology*, *21*(10), 640–656. DOI: 10.1038/s41579-023-00900-7

Singh, J., Sahany, S., Singh, K. K., Robock, A. and Xia, L. (2024). Future climate change impacts on rice in Uttar Pradesh, India's most populous agrarian state. *Earth's Future*, 12(5), e2023EF004009.

Singh, R. P., Kumar, S., Sainger, M., Sainger, P. A. and Barnawal, D. (2017). Eco-friendly nitrogen fertilizers for sustainable agriculture. *Adaptive soil management: From theory to practices*, 227-246.

Sivakumar, M. V. (2007). Interactions between climate and desertification. *Agricultural and Forest Meteorology*, *142*(2-4), 143–155. DOI: 10.1016/j.agrformet.2006.03.025

Skendžić, S., Zovko, M., Živković, I. P., Lešić, V., & Lemić, D. (2021). The impact of climate change on agricultural insect pests. *Insects*, *12*(5), 440. DOI: 10.3390/insects12050440

Smilovic, M., Gleeson, T., & Siebert, S. (2015). The limits of increasing food production with irrigation in India. *Food Security*, *7*(4), 835–856. DOI: 10.1007/s12571-015-0477-2

Somasundaram, J., Sinha, N. K., Dalal, R. C., Lal, R., Mohanty, M., Naorem, A. K., Hati, K. M., Chaudhary, R. S., Biswas, A. K., Patra, A. K., & Chaudhari, S. K. (2020). No-till farming and conservation agriculture in South Asia–issues, challenges, prospects and benefits. *Critical Reviews in Plant Sciences*, *39*(3), 236–279. DOI: 10.1080/07352689.2020.1782069

Stocker, T. F., Qin, D., Plattner, G. K., Alexander, L. V., Allen, S. K., Bindoff, N. L., Bréon, F. M., Church, J. A., Cubasch, U., Emori, S., & Forster, P. (2013). Technical summary. In *Climate change 2013: the physical science basis. Contribution of Working Group I to the Fifth Assessment Report of the Intergovernmental Panel on Climate Change* (pp. 33–115). Cambridge University Press.

Subash, N., Gangwar, B., Singh, R., & Sikka, A. K. (2015). Identification of climate-resilient integrated nutrient management practices for rice-rice cropping system—An empirical approach to uphold food security. *International Journal of Biometeorology*, *59*(1), 65–78. DOI: 10.1007/s00484-014-0825-5

Taheripour, F., Hertel, T. W., Gopalakrishnan, B. N., Sahin, S., & Escurra, J. J. (2015). Agricultural production, irrigation, climate change, and water scarcity in India.

UNFCCC. 2017. Glossary of climate change acronyms and terms.

Verge, X. P. C., De Kimpe, C., & Desjardins, R. L. (2007). Agricultural production, greenhouse gas emissions and mitigation potential. *Agricultural and Forest Meteorology*, *142*(2-4), 255–269. DOI: 10.1016/j.agrformet.2006.06.011

Vivekanandan, E., Rajagopalan, M., & Pillai, N. G. K. (2009). Recent trends in sea surface temperature and its impact on oil sardine.

Wasti, S., Sah, N., & Mishra, B. (2020). Impact of heat stress on poultry health and performances and potential mitigation strategies. *Animals (Basel)*, *10*(8), 1266. DOI: 10.3390/ani10081266

Wu, W., & Ma, B. (2015). Integrated nutrient management (INM) for sustaining crop productivity and reducing environmental impact: A review. *The Science of the Total Environment*, *512*, 415–427. DOI: 10.1016/j.scitotenv.2014.12.101

Zhang, F., Cui, Z., Chen, X., Ju, X., Shen, J., Chen, Q., Liu, X., Zhang, W., Mi, G., Fan, M., & Jiang, R. (2012). Integrated nutrient management for food security and environmental quality in China. *Advances in Agronomy*, *116*, 1–40. DOI: 10.1016/B978-0-12-394277-7.00001-4

Zhao, Q., & Dixon, R. A. (2014). Altering the cell wall and its impact on plant disease: From forage to bioenergy. *Annual Review of Phytopathology*, *52*(1), 69–91. DOI: 10.1146/annurev-phyto-082712-102237

Chapter 2
Use of Artificial Intelligence and Information and Communication Technology for Efficient Nutrient and Water Management in Agriculture

Varsha Rattan
CSK Himachal Pradesh Krishi Vishvavidalaya, India

Narender Kumar Sankhyan
https://orcid.org/0009-0007-4297-825X
CSK Himachal Pradesh Krishi Vishvavidalaya, India

Sanjeev Kumar Sandal
CSK Himachal Pradesh Krishi Vishvavidalaya, India

DOI: 10.4018/979-8-3693-4392-0.ch002

ABSTRACT

As the global population approaches 9.7 billion by 2050, sectors such as agriculture and water supply are facing unprecedented challenges. To address these pressures, enhancing resource resilience and sustainability is crucial. Technological innovations in AI and ICT offer a promising solution. Tools like Conserwater, SupPlant, Plantix, and advanced smart irrigation systems are revolutionizing how we manage resources. These technologies provide real-time monitoring and decision support, optimizing efficiency. By leveraging AI techniques such as machine learning and deep learning, integrated with GIS, smart meters, and mobile devices, we gain predictive insights that significantly improve nutrient and water management. This technological shift enables more sustainable agricultural practices, helping us use resources more effectively and meet the demands of a growing global population.

INTRODUCTION

Water and nutrients are essential elements that sustain all forms of life, including both plants and humans. They are the foundation for the growth and development of living organisms, playing a crucial role in maintaining health and vitality. In the context of agriculture, the efficient management of water and nutrients is directly linked to food and nutritional security, which is vital not only for the well-being of current populations but also for ensuring the sustainability and prosperity of future generations (FAO, 2019). Every form of life on our planet is intricately dependent on these vital resources. For plants, water and nutrients are indispensable; they play crucial roles in various processes that support plant health and productivity. These resources are essential for structural support, nutrient uptake and transportation, biochemical reactions, and energy provision within the plant system. Despite their apparent abundance, the accessibility of water and nutrients is not uniform across the globe. While some regions enjoy relatively easy access to these resources, others face severe shortages that can have life-or-death implications. The rapid increase in the global population has resulted in a sharp rise in agricultural production to meet the growing demand for food. This expansion in food production has, however, put tremendous strain on vital water resources, leading to their overuse and depletion. Additionally, the intensified agricultural activities have caused the accelerated removal of key nutrients from the soil, further threatening the long-term fertility and sustainability of the land. These challenges highlight the need for more sustainable agricultural practices to balance production and environmental conservation

The scarcity and quality of water and nutrients present significant challenges that have far-reaching implications for both food security and environmental sustainability. With the growing global population and increasing demand for food, the need for effective management of water and nutrients has become more critical than ever. These resources are not only vital for agricultural productivity but also for the health and well-being of ecosystems. To effectively address these challenges, there must be a concerted effort from all stakeholders—including governments, agricultural industries and research institutions—to improve the management of these critical resources in a sustainable manner. This involves adopting innovative and holistic approaches that not only ensure the efficient use of water and nutrients but also protect the environment from degradation. Efficient water and nutrient management practices in agriculture are essential for mitigating issues such as water scarcity, soil nutrient depletion, and land degradation. In many regions, especially those experiencing arid and semi-arid conditions, water scarcity poses a serious threat to agricultural productivity. Moreover, the overuse or mismanagement of water resources can lead to the contamination of freshwater supplies, affecting both crop production and human health. Similarly, nutrient depletion, particularly the loss of essential elements like nitrogen, phosphorus, and potassium from the soil, can lead to reduced crop yields and lower food quality. Addressing these issues requires a comprehensive, multi-faceted approach that tackles specific problems such as water scarcity, contamination, wastage and nutrient depletion, while simultaneously working towards the restoration and conservation of natural ecosystems.

A key component of this integrated approach involves the adoption and implementation of advanced technologies, such as artificial intelligence (AI) and information and communication technologies (ICT), which are revolutionizing the field of agriculture. These state-of-the-art tools are enabling the development of a digitally-empowered farming system that enhances resource use efficiency and reduces waste to a minimum. AI and ICT are helping farmers make data-driven decisions by providing real-time information on water usage, soil health, crop conditions and weather patterns. This allows for precise irrigation and fertilization, optimizing water and nutrient use and reducing the risk of overuse or waste. For instance, AI-powered algorithms can predict crop water requirements based on weather forecasts and soil moisture data, while sensors and drones can provide detailed information on nutrient deficiencies in the soil, enabling targeted interventions. Furthermore, the integration of these technologies can help improve agricultural resilience to climate change, which is expected to exacerbate challenges related to water scarcity and nutrient availability. By providing accurate, timely data, AI and ICT can help farmers adapt to changing environmental conditions, making their practices more sustainable in the long term. These technologies can also contribute to reducing the carbon footprint of agriculture by optimizing inputs and minimizing waste, ultimately promoting environmental sustainability.

In addition to technology, effective water and nutrient management also requires a shift in agricultural policies, education and public awareness. Governments must implement policies that incentivize sustainable practices, provide access to modern farming tools, and promote the adoption of water- and nutrient-efficient technologies. Furthermore, educating farmers about the benefits of efficient resource management, as well as providing them with the necessary training and support, is essential for the successful implementation of these practices. Local communities, agricultural extension services and research institutions must collaborate to provide farmers with the knowledge and resources needed to improve their agricultural practices and ensure the sustainable use of water and nutrients. Addressing the challenges of water scarcity and nutrient depletion requires a multi-dimensional approach that integrates advanced technologies, policy support, education and sustainable farming practices. By enhancing the efficiency of water and nutrient use and minimizing waste, agriculture can meet the growing demand for food while ensuring environmental sustainability. The integration of AI and ICT in farming systems is a promising step towards achieving these goals, helping to create a more resilient and sustainable agricultural sector that can thrive in the face of changing global conditions.

The convergence of Artificial Intelligence (AI) and Information and Communication Technologies (ICT) in agriculture is shaping the future of farming in profound ways. These technologies are not only enhancing the efficiency of agricultural practices but also driving the transition toward more sustainable and resource-conscious systems. By integrating AI and ICT, we can create innovative solutions that tackle some of the most pressing challenges facing agriculture today, such as water scarcity and nutrient management. Precision farming, enabled by these technologies, allows for more accurate monitoring and optimization of water and soil resources, reducing waste and improving crop yields. Furthermore, AI-driven systems can predict weather patterns, disease outbreaks, and pest infestations, helping farmers make more informed decisions and reduce dependency on harmful chemicals. This digital transformation is crucial for developing resilient food systems that can meet the demands of a growing global population. By improving resource use efficiency and minimizing environmental impacts, AI and ICT are fostering greater environmental sustainability. Ultimately, these advancements contribute to the creation of a more secure, sustainable, and equitable agricultural framework, ensuring healthier food systems that benefit both people and the planet.

1. ARTIFICIAL INTELLIGENCE (AI)

Due to the expansion in population, urbanization and climate change, competition for resources is increasing with the passage of time in order to meet the world's increasing food demand. The increased demand for food contributes to expanding cultivation to a great extent and excessive pumping of large volume of water for irrigation purpose and mismanagement of water has resulted in soil productivity loss and shortage of water resource, therefore improving nutrient and water management is the urgent necessity which AI is a great boon to the agriculture sector as it is highly dependent on climatic conditions which is unpredictable and aids to enhance the productivity and resource management. It helps automate simple and mundane tasks such as gathering field information, labelling and analyzing data, developing reports and sending notifications. AI technologies plays a vital role in water and nutrient management as it manages by designing efficient supply systems, optimizing present resources, planning infrastructure for these resources, which involve use of software powered programs with sensors and neural networks that strategize various operations dynamically and intelligently (Doorn, 2021).

Artificial intelligence (AI) represents an advanced and revolutionary tool designed to emulate human intelligence through machine processes. This advanced technology predominantly operates through computer systems, robotics and various digital devices. AI is fundamentally comprised of three key components: data input, algorithms and output. Data input encompasses a broad range of formats, including time series data such as rainfall patterns, historical yield data from farmers and visual inputs like images. These inputs provide the raw information that AI systems analyze and process. The term "algorithm" refers to a systematic series of calculations or instructions that AI systems follow to address and solve specific problems. These algorithms are crucial as they guide the processing and analysis of the data. They involve various computational techniques and methods to transform input data into actionable insights. AI algorithms facilitate complex data processing, automate reasoning, and perform advanced calculations, allowing the system to derive meaningful outcomes from the given data. The output of AI systems consists of the results generated after processing the input data through algorithms. These outputs can take various forms, including recommendations, predictions, and classifications. For instance, AI might provide forecasts for anticipated rainfall in the coming year, project agricultural yields, or identify and categorize objects detected within images. The essence of AI solutions lies in their ability to deliver smart recommendations and precise predictions based on the data they analyze.

AI aims to offer valuable insights and actionable information by leveraging data inputs, sophisticated algorithms and advanced output mechanisms. Through these elements, AI systems are able to simulate human-like intelligence and provide users with informed, data-driven decisions across numerous applications. This innovative technology continues to evolve, enhancing its capabilities to address complex problems and deliver increasingly accurate and useful predictions.

Figure 1.

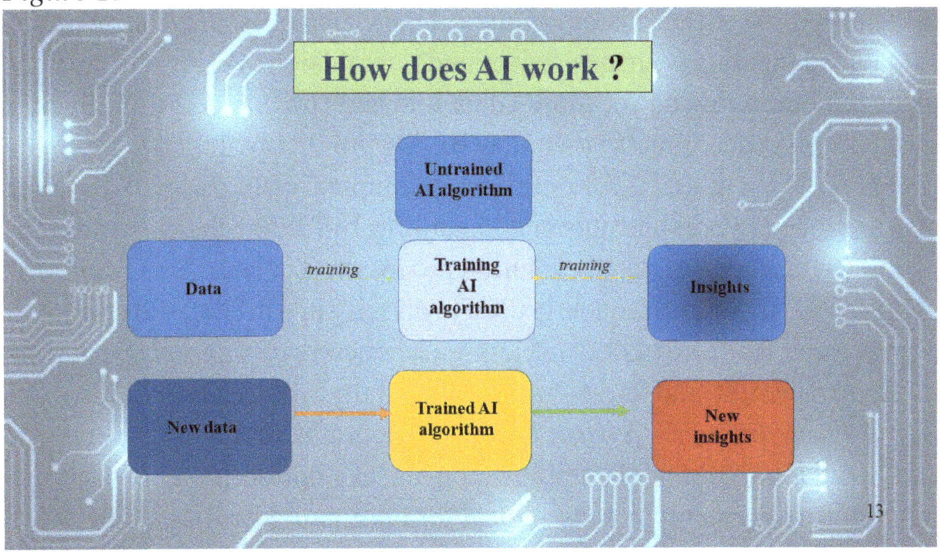

AI systems works by assimilating large amounts of labelled training data such as weather data, analyzing the data for correlations and patterns, and using these patterns to make predictions or weather forecasts which are helpful for irrigation scheduling or carrying out other operations. It mainly focuses on three cognitive skills which are learning, reasoning and self correction. In learning process, all the available information is acquired and certain rules are created using the algorithm that provides computing devices with instructions to complete a specific task. The reasoning process emphasizes on the right algorithm to reach a desired outcome and self correction aspect of AI programming is designed to continually fine-tune algorithms and ensure to provide the most accurate results possible. AI holds within two sub-fields that are machine learning and deep learning, which are frequently mentioned in conjunction with artificial intelligence. These disciplines are comprised of AI algorithms which seek to create expert systems that make predictions or classifications based on input data.

Several advanced models within artificial intelligence (AI) are extensively used in the water and nutrient management sector to improve decision-making, resource allocation, and overall efficiency. These models include *Artificial Neural Networks (ANN)*, which are applied to predict water and nutrient requirements, optimize irrigation systems and manage crop growth based on historical data. *Fuzzy Logic Algorithms (FL)* are valuable for dealing with uncertainty and imprecision in water and nutrient management, allowing for more adaptable and flexible decision-making in situations where precise data may be difficult to obtain. *Genetic Algorithms (GA)* are utilized for optimizing complex systems, such as water distribution networks and nutrient scheduling, by finding the most efficient solutions through simulated evolution processes. The *Support Vector Machine (SVM)* model is frequently used to classify and predict water-related events, such as droughts or nutrient deficiencies, by analyzing patterns in environmental and crop data. Additionally, the *Adaptive Network-Based Fuzzy Inference System (ANFIS)* integrates both neural networks and fuzzy logic to model complex, non-linear relationships in water and nutrient management, making it ideal for optimizing irrigation schedules and nutrient delivery systems. Together, these AI-driven models help to enhance efficiency, improve sustainability and optimize the management of water and nutrient resources in agriculture.

Some AI based technologies for nutrient and water management:

Plantix: This advanced crop advisory application leverages the power of artificial intelligence (AI) to provide comprehensive support for farmers, helping them identify and address a wide range of crop-related issues. Combining cutting-edge deep learning algorithms and sophisticated image recognition technology with the expertise of leading agricultural research institutions, the system offers a highly effective, data-driven solution. The application analyzes images of crops to detect a variety of problems, including nutrient deficiencies, plant diseases, and pest infestations, all of which can affect crop health and yield. By employing deep learning models, the system processes and interprets the visual data, enabling it to detect issues that may not be easily visible to the human eye. In addition to image analysis, the application also gathers and evaluates data on the spatial distribution of crops, which further enhances its ability to deliver precise, context-specific diagnostics. Based on the insights derived from this data, the application generates tailored recommendations for farmers, offering a range of management practices and intervention strategies to address the identified problems. These recommendations include both preventive measures, such as adjustments to irrigation or fertilization, and corrective actions, such as targeted pest control or disease management techniques. By receiving these personalized, actionable insights, farmers are better equipped to manage their crops effectively, improve crop health, and boost productivity. Ultimately, the integration of AI with expert knowledge empowers farmers to make informed decisions that lead to higher yields, reduced resource waste and more sustainable agricultural practices.

Figure 2.

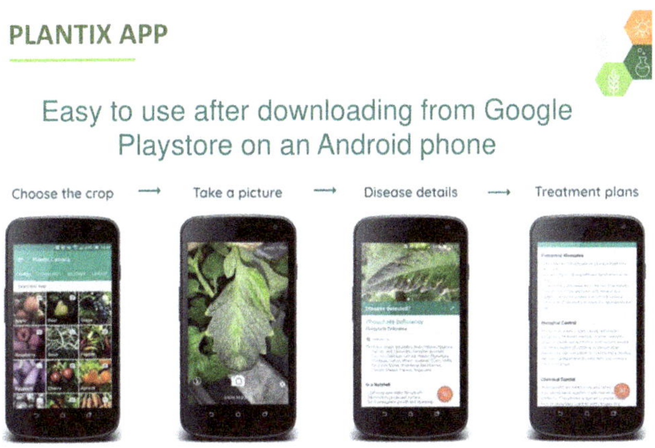

IBM Agropads: Agropad is an *innovative* technology designed to offer advanced and precise monitoring of soil and water health. This state-of-the-art device allows for real-time, on-site chemical analysis of soil or water samples, utilizing the power of artificial intelligence to enhance its accuracy and efficiency. The Agropad system consists of a specialized card, with several circles on its reverse side, each containing a microfluidic chip that conducts detailed chemical analyses. These microfluidic chips are meticulously calibrated to detect a wide variety of chemicals, providing a comprehensive understanding of the chemical makeup of the sample being analyzed. The real strength of Agropad lies in its ability to perform accurate chemical analyses within minutes, directly at the site where the sample is collected. After the analysis is completed, the device employs advanced machine learning algorithms and machine vision techniques to process the color intensity data gathered from the microfluidic chip. These algorithms interpret the color changes, which correspond to specific chemical concentrations, transforming the raw data into clear and actionable results. This process ensures that the results provided are not only precise but also immediately useful to the user, allowing for prompt decision-making. The integration of artificial intelligence and machine learning further amplifies Agropad's effectiveness, as the system continuously learns and adapts based on previous data, improving its accuracy over time. This cutting-edge technology ensures that users can receive highly reliable, real-time assessments of soil and water quality. The ability to rapidly analyze chemical composition without needing laboratory facilities makes Agropad a highly convenient and accessible solution for farmers, environmentalists, and researchers working in the field.

In agricultural settings, this tool enables farmers to optimize irrigation practices, monitor soil health, and make data-driven decisions about crop management. In environmental contexts, Agropad aids in water quality monitoring and pollution control efforts, making it an invaluable tool for both sustainable farming and environmental protection. By offering instant, accurate insights, Agropad empowers users to take immediate, informed action, leading to improved agricultural productivity, resource conservation, and better environmental stewardship. This remarkable combination of speed, accuracy, and user-friendliness makes Agropad a transformative tool for enhancing soil and water management practices globally.

Figure 3.

Figure 4.

Trace genomics: Trace Genomics represents a groundbreaking technological approach that combines the fields of soil science, genomics, and machine learning to conduct in-depth analyses of soil biology and chemistry. By merging these advanced disciplines, this innovative technology provides a far more comprehensive understanding of soil health and its functionality, offering farmers and researchers valuable insights into the complex dynamics of the soil ecosystem. Trace Genomics uses sophisticated machine learning algorithms alongside genomic data to assess the interactions between various soil microorganisms, their biochemical processes, and the soil's overall fertility. This integration of soil science and genomics enables the identification of unique microbial communities that play a critical role in maintaining soil fertility, enhancing nutrient cycling, and promoting plant growth. By uncovering these intricate relationships, Trace genomics helps to identify potential issues within the soil that could hinder crop productivity, such as nutrient imbalances or harmful microbial populations. Armed with this data, farmers can implement targeted, science-driven interventions to optimize soil health and fertility. Ultimately, Trace genomics offers novel, data-driven solutions for sustainable agriculture, contributing to the preservation of soil health and maximizing its potential. Through better soil management practices, this technology supports the improvement of crop yields, enhances soil productivity, and encourages environmentally responsible farming, making it a key tool for advancing sustainable agricultural practices globally.

Conserwater: Conserwater is an innovative water conservation platform designed to optimize irrigation practices through the integration of satellite imagery, real-time weather data, and advanced artificial intelligence algorithms, including deep learning techniques. This cutting-edge technology operates without requiring physical hardware, relying on data from a variety of sources to predict irrigation needs with remarkable accuracy. By combining these data streams, Conserwater provides an extremely precise assessment of a crop's water requirements, enabling farmers to make well-informed decisions about when and how much to irrigate. The platform's ability to analyze these complex data sets ensures that water usage is optimized, leading to significant reductions in the amount of water, electricity, and fertilizers required for crop cultivation. Furthermore, this results in lower operational costs, allowing farmers to allocate resources more efficiently. In contrast to traditional evapotranspiration-based irrigation systems, which rely on standard formulas and physical soil sensors, Conserwater offers a highly effective and cost-efficient alternative. The platform has been shown to reduce irrigation water usage by approximately 30per cent or more when compared to conventional methods, making it an invaluable tool for modern farming. This impressive water-saving efficiency not only helps farmers achieve better crop yields but also plays a crucial role in minimizing environmental impact by conserving valuable water resources. By providing farmers with data-driven, actionable insights, Conserwater ensures that irrigation practices are more effective, sustainable, and environmentally friendly. As a result, it becomes a key asset for farmers seeking to improve productivity while contributing to the preservation of natural resources. The platform's role in advancing sustainable agriculture and promoting water conservation makes it a vital tool in addressing the global challenges of resource scarcity and environmental sustainability.

SupPlant: This innovative AI-powered irrigation system utilizes advanced algorithms to process real-time data collected from a wide range of sensors that monitor key factors such as plant health, soil conditions, and meteorological variables. By gathering data from multiple sources, the system creates a comprehensive understanding of the growing environment, which is then translated into precise irrigation recommendations and actionable commands. The sensors used in this system include dendrometers, which measure the diameter of plant stems or bark to assess growth, fruit growth sensors that monitor the development of crops, and subminiature touch probes that measure the temperature of leaves, providing critical information on plant stress and water needs. Designed to facilitate automated, growth-based irrigation, this cutting-edge technology enables farmers to adjust watering schedules and amounts based on the real-time needs of the plants, optimizing water usage while ensuring optimal crop health. The system has proven highly effective, reducing water consumption by approximately 20 per cent, while simultaneously increasing crop yields by about 10 percent. By integrating AI-driven analytics with state-of-the-art sensor technology, this system significantly improves water management, making it a powerful tool for enhancing agricultural productivity and promoting sustainable farming practices. Ultimately, it helps farmers use water resources more efficiently, contributing to both environmental sustainability and improved agricultural outputs.

Digital farming: Developed in Japan, this cutting-edge technology combines the power of artificial intelligence (AI) and the Internet of Things (IoT) to revolutionize the fertigation process, offering farmers a more efficient, precise and sustainable method of managing water and fertilizer application. The system works by gathering real-time data from a range of sensors that measure environmental factors such as light intensity, soil moisture, temperature and nutrient levels. This data is then processed and analyzed using sophisticated AI algorithms, which also incorporate the accumulated knowledge and best practices from skilled cultivators and agricultural experts. The result is a highly personalized approach to fertigation, where the system provides tailored recommendations for the precise amounts of water and fertilizers needed for optimal crop growth. This allows farmers to address the unique needs of their crops, enhancing growth and improving yields. One of the most significant advantages of this technology is its ability to help farmers overcome common challenges such as water scarcity. By optimizing the use of water and fertilizers, the system reduces overall consumption, cutting down on waste and improving resource efficiency. Furthermore, the system is designed to support both experienced and novice farmers. For those with less expertise, the AI-driven recommendations provide valuable guidance, making the technology accessible and easy to use. This leads to improved productivity and better outcomes, even for those new to farming. By ensuring that resources are used more effectively and minimizing the environmental impact, the technology promotes more sustainable farming practices. It helps farmers not only maximize their crop yields but also contribute to the long-term health of the environment and the sustainability of agricultural systems worldwide.

Bluetentacles: This precision irrigation solution, grounded in advanced AI technology, is designed to optimize water usage by applying it only when necessary. The system is exceptionally user-friendly, tailored specifically to address the needs of farmers. It features a straightforward interface where Internet of Things (IoT) devices continuously monitor the field and gather relevant data. This data is then transmitted to the cloud via LoRaWAN (Long Range Wide Area Network). Once in the cloud, the data is analyzed and integrated with additional information, such as meteorological data. Based on this comprehensive analysis, precision irrigation recommendations are generated and sent back to the device, which is directly connected to hydro valves. This process ensures that water is applied efficiently and accurately, enhancing overall irrigation management and resource conservation.These AI-based technologies mark a significant advancement in enhancing nutrient and water efficiency across various agricultural settings, including both open fields and greenhouses. The exceptional accuracy, precision, and reliable, error-free performance of these AI systems have played a pivotal role in their increasing adoption within the agricultural sector. Their capability to continually optimize resource use and improve farming practices demonstrates their transformative impact on modern agriculture. By enabling more precise management of nutrients and water, these technologies contribute to greater efficiency and sustainability in farming operations. (Malviya & Jaspal, 2021).

2. INFORMATION AND COMMUNICATION TECHNOLOGY (ICT)

The integration of Information and Communication Technology (ICT) plays a pivotal role in enhancing the management of both nutrients and water resources. It enables the efficient generation, storage, analysis, dissemination, and application of vast amounts of data, thereby improving agricultural practices. With ICT, farmers and agricultural professionals gain access to timely, accurate, and location-specific information, allowing for more informed decision-making. This technology facilitates the optimization of resources, leading to better water and nutrient management, which in turn boosts production efficiency. By providing targeted insights, ICT is instrumental in enhancing both productivity and sustainability, making it an essential tool for advancing resource management practices in modern agriculture (Singh et al., 2017).

In the agricultural sector, as well as in urban water supply and industrial applications, ICT plays a crucial role in advancing nutrient and water use efficiency. It addresses critical challenges such as soil degradation, water scarcity, and vulnerabilities within agriculture. ICT-based technologies, including software and sensors, facilitate the automation of monitoring processes and the sharing of databases, which leads to improved service delivery with comprehensive nationwide coverage. This automation supports better performance benchmarks for nutrient and water use, promoting greater accountability and transparency within the agriculture sector. Through these advancements, ICT not only optimizes resource management but also fosters sustainable practices and enhances overall operational effectiveness.

There are various ICT tools like cloud computing, sensors, internet, smartphones, remote sensing, satellite imagery, smart meters, geographic information system, various software like Decision support system (DSS) which provides daily information based on the need of the farmers and helps to understand the gap between the productivity and market demand. (Das et al. 2017). Implementation of various ICT based solution/tools such as sensors, geographic information system, remote sensing & advanced communication tools has brought about real-time data base and information management and thus improving the water utilities. Smart meters, sensors and smart phones are much likely to benefit by promoting efficient use of nutrient and water, both at the supply and demand side through monitoring of total consumption, smart billing system, pump efficiency, prevention of potential problems such as slow leakage, clogged drain or over flow of water ensure water conservation and reduction losses.

Barkunan et al. (2019) studied the use of sensor for automatic drip irrigation for paddy and noticed that the proposed smart sensor based irrigation system saved nearly 41.5 per cent and 13 per cent of water compared to the conventional flood and drip irrigation methods respectively. It helps in scheduling the optimal time of irrigation along with optimal water use for agriculture, which helps in preventing damage due to drought stress or over irrigational practices. Singh et al. (2015) conducted a field experiment in Ludhiana, Punjab, to investigate sensor-based nitrogen management in a rice cultivar during the June to September period. The results of the study indicated that the treatment in which nitrogen was applied based on guidance from the Green Seeker sensor resulted in significantly higher grain yield and improved nitrogen use efficiency. This approach demonstrated its effectiveness in optimizing nitrogen application, leading to enhanced crop productivity and more efficient use of fertilizers. Through the use of various ICT tools like smart phones, internet or GPS the remote management of irrigation activities, fertilizer usage, extracting groundwater from pumps can be done and these tools also help the water authorities in regulating allocation of irrigation water from reservoirs and minimizing the distribution losses in command areas (Anonymous, 2019). Singh et al. (2021) conducted a study on sensor-based and decision support system-driven irrigation for maize and soybean crops under sub-humid conditions. Their findings revealed that the implementation of these advanced irrigation technologies resulted in significantly higher irrigation water use efficiency for both crops. The sensor and decision support system not only optimized water usage but also contributed to improved resource management, highlighting the effectiveness of these systems in enhancing irrigation practices for better crop productivity and water conservation.

ICT Based Platforms for Water and Nutrient Management

Smart irrigation systems: This advanced system combines a variety of sensors and automation technologies to efficiently monitor and regulate the water needs of plants. By integrating technologies such as automation platforms like Arduino or Raspberry Pi, along with a range of sensors—solar sensors, soil moisture sensors, temperature sensors, and humidity sensors—the system continuously collects data to assess the specific requirements of each plant. Artificial intelligence (AI) is then used to analyze this data and make accurate decisions about the optimal irrigation levels, ensuring that each plant receives the precise amount of water it needs for healthy growth, while conserving resources. The result is an automated irrigation system that can operate independently, without requiring manual intervention from an operator. This system ensures that each plant receives the exact amount of water needed, preventing issues like dehydration, over-irrigation, or irregular watering, which could potentially harm the plants and lead to diseases. A key feature of this system is the ability to exchange remote information between the system and the user, acting as a feedback loop. This feedback mechanism helps ensure the system remains efficient, offering a nearly fail-safe approach to autonomous irrigation.

According to Kamil and Ojoawa (2019), such a setup is highly reliable and can improve water management in agricultural settings.

Several experiments have demonstrated the effectiveness of smart irrigation systems. For example, a study using machine learning techniques in a smart irrigation decision support system resulted in a 22 per cent reduction in irrigation requirements for citrus crops, significantly conserving water resources (Navarro et al., 2016). In another study, Liu and Xu (2018) investigated an automatic irrigation control system for soilless lettuce cultivation during the spring and summer months. They found that the system achieved irrigation water use efficiency rates of 68.03 per cent in spring and 98.61 per cent in summer, both significantly higher than traditional manual irrigation methods. Additionally, the system produced a lower drainage rate compared to manual control irrigation. Similarly, Mohammad et al. (2013) worked on an intelligent irrigation system for tomato crops and observed that both crop yield and water use efficiency were higher than those achieved with conventional irrigation control systems. These studies highlight the substantial benefits of intelligent, automated irrigation systems in terms of water conservation, efficiency and crop productivity.

Figure 5.

Figure 6.

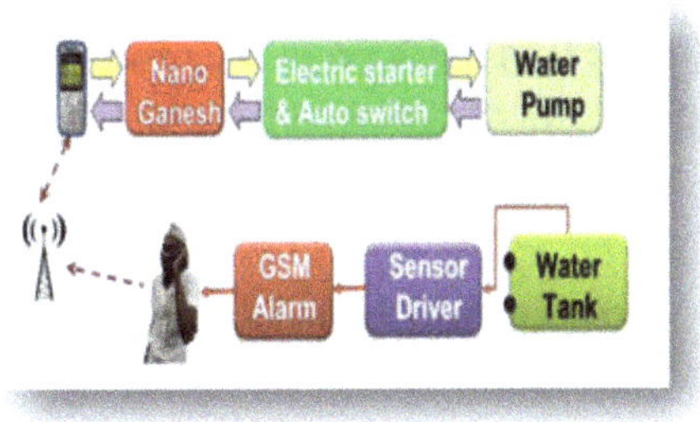

One of the smart irrigation system is cloud and solar hybrid smart irrigation system (CS-HSIS) in which controller unit, wireless medium, GSM module, arduino, display unit, Raspberry Pi, water pump, smart meters, sprinklers/drip module is used. In this system power is supplied through solar panel and its consists of various sensors like temperature sensor, soil moisture sensor and humidity sensor which are directly connected to arduino controller that is responsible to turn the water pump on and off when required. A concept of existing knowledge (from previous weather data) through raspberry pi plays a very significant role as it prevents the plants from getting dehydrated when the sensors are not working. This system is also able to send all irrigation-related data to the cloud, for making analysis and future purposes. User can see the current and past irrigation usages anytime on a smart web-based or app-based portal. Smart valve/ smart meter will count the units of water consumptions and send these details to the cloud and accordingly the wifi enabled sprinklers provide irrigation keeping in check the speed of water and the direction of the nozzles. This enables to know the total water consumption/requirements for a particular crop and monthly/weekly consumption of water. These values can be used for making future plans/decisions about water requirements (Sinwar et al. 2019).

Miller et al. (2018) compared smartphone irrigation scheduling application (VegApp) with water balance (WB) and soil moisture sensor-based (SMS) irrigation methods in tomato crop and noticed that the VegApp- and WB-based methods applied less water on a daily basis in 2017 than 2016 because of lower ETc and SMS based had the most consistent irrigation water use efficiency in both the years. A smartphone application, *Blueleaf*, was evaluated for its effectiveness in irrigation scheduling for durum wheat, comparing its performance to traditional farmer practices and rainfed conditions. The study revealed that the use of the Blueleaf application resulted in a significant increase in biomass production, with the crop growing 41.76per cent more biomass than the rainfed crop. Similarly, the biomass yield from the farm managed with the Blueleaf app was 41.37per cent higher than the rainfed crop. In terms of grain yield, Blueleaf outperformed rainfed conditions by a remarkable 67.14 per cent, while the farm using the Blueleaf app achieved a 67.50 per cent higher grain yield compared to the rainfed crop. Notably, despite the higher yields, the volume of water used for irrigation was significantly lower under the Blueleaf treatment, indicating a more efficient use of water resources. This demonstrates the potential of the Blueleaf application in improving crop productivity while optimizing water usage, making it a valuable tool for sustainable farming practices (Saab et al., 2019).

Water productivity open-access portal (WaPOR): The FAO has introduced a pioneering and sophisticated portal that integrates advanced remote sensing technology to revolutionize the monitoring and evaluation of agricultural water productivity. This advanced tool is designed to provide an extensive suite of features, empowering users to engage with large-scale datasets, conduct comprehensive time series analyses and generate precise, area-specific statistics related to water and land use. The portal serves as a critical resource for countries by offering detailed insights into their water productivity levels. It facilitates the identification and thorough analysis of productivity gaps, enabling users to understand where inefficiencies lie and how they impact agricultural performance. This detailed examination supports a better grasp of water and land use patterns, making it possible to detect areas requiring intervention and improvement. Beyond merely identifying gaps, the portal is equipped with capabilities to propose and implement strategic solutions aimed at closing these gaps. By leveraging its analytical power, users can develop targeted interventions to optimize water management and enhance agricultural efficiency. This, in turn, leads to more sustainable agricultural practices and more effective resource utilization. Moreover, the portal's comprehensive data and analytical tools assist in crafting policies and strategies that support sustainable agricultural development. By offering actionable insights and facilitating the formulation of data-driven solutions, it plays a vital role in driving improvements in agricultural production. The ultimate goal of the portal is to contribute to a significant and enduring increase in agricultural output while promoting efficient and sustainable water management practices.

Sustainable Water and Innovative Irrigation Management (SWIIM): This sophisticated software suite and technology system is designed to offer advanced irrigation monitoring solutions that enhance farm operations and conserve applied agricultural water, ultimately boosting the overall potential income of farming enterprises. The system comprises two integral modules: SWIIM Planner and SWIIM Manager. The SWIIM Planner is tailored for strategic water management, allowing users to develop and implement effective irrigation plans. It facilitates precise planning by considering various factors to ensure optimal water use. On the other hand, the SWIIM Manager focuses on the management of irrigation practices, helping users monitor and adjust their strategies in real time. The software suite provides comprehensive, customized reports that include detailed information on water delivery, water balance and conservation practices. It also leverages remote-sensing technology to offer valuable insights into water use optimization, water rights management and potential increases in net income from farm operations. By integrating these advanced tools and technologies, the system not only ensures efficient water conservation but also significantly boosts overall farm productivity and profitability. It helps farmers optimize resource use, minimize waste, and reduce costs, thereby supporting more sustainable agricultural practices. Furthermore, the system contributes to improved operational outcomes, allowing farms to increase their output while promoting environmental stewardship and long-term viability in agricultural operations.

Nanoganesh: Nano Ganesh is an innovative and groundbreaking mobile phone-based remote control system designed to enable farmers to manage and monitor their water pumps from a distance. This user-friendly system provides a highly convenient and efficient solution for water management, which is essential for agricultural success. Developed by engineer Santosh Ostwal, Nano Ganesh bridges the gap between high-voltage electrical starters and low-voltage mobile devices, making it accessible and easy for farmers from various backgrounds to use. The system allows farmers to control their water pumps remotely, providing them with the ability to turn the pumps on or off directly from their mobile phones, even if they are located far away from the pump site. In addition to controlling the pumps, Nano Ganesh also provides farmers with the ability to check the status of the water supply near the pump. This feature ensures that water is being supplied efficiently and allows farmers to manage their water resources more effectively. The technology is the culmination of nearly two decades of development, carefully crafted to meet the needs of farmers with varying levels of education, language preferences, age and gender. Nano Ganesh is specifically designed to be simple and intuitive, empowering farmers to take full control of their irrigation systems without requiring extensive technical knowledge. The application of Nano Ganesh technology is particularly beneficial in helping farmers plan their water usage more effectively, reducing water wastage and optimizing the use of water resources. This leads to more sustainable agricultural practices, helping to conserve water, reduce energy consumption and improve crop yields. By enhancing water management efficiency, Nano Ganesh contributes significantly to sustainable farming practices, making it a valuable tool in modern irrigation management. The simplicity, affordability and ease of use of Nano Ganesh have made it a transformative solution for farmers, increasing productivity while promoting resource conservation and a more sustainable approach to agriculture.

AQUASTAT: The Food and Agriculture Organization of the United Nations (FAO) has developed a comprehensive global information system that focuses on water resources and agricultural water management, addressing one of the most critical challenges for food security and environmental sustainability worldwide. This system plays an integral role in improving the management of water resources, particularly in developing countries and countries in transition, where water scarcity and inefficient use of water resources are prevalent. The FAO's platform consists of several key components that provide essential data and insights for decision-making, research and policy development.A core aspect of the system is its systematic descriptions of the state of agricultural water management across different countries and regions. This information highlights the current state of water management practices, with a special focus on regions where water management practices need to be improved. It emphasizes the challenges faced by developing nations in managing water resources effectively, with agriculture being a major consumer of water. The system also provides up-to-date online data by country, offering real-time statistics on water availability, usage, and agricultural water management, enabling stakeholders to make informed decisions based on the most current information. In addition, the platform includes digital geographical data on water resources and irrigation systems, such as maps and satellite imagery, which provide valuable tools for understanding water distribution and optimizing irrigation practices. Specific studies, including reviews of world water resources by country and assessments of the irrigation potential in Africa, further enhance the understanding of global water challenges and opportunities for irrigation development. The system also offers projections of future agricultural water use and irrigation expansion, helping policymakers plan for future water demand.

Phule irrigation scheduler: This mobile or web-based application has been designed to estimate the reference evapotranspiration (ETr) by utilizing weather data inputs such as maximum temperature, minimum temperature, maximum relative humidity, minimum relative humidity, wind speed, and sunshine hours, sourced from the "Open Weather" service provider. The application provides real-time weather information for the current day and forecasts for the next three days. Additionally, the app calculates crop-specific coefficient values (Kc) on a daily basis, which are determined based on crop type and growth stage. These crop coefficients are derived from a combination of literature and experimental data collected as part of this project. To estimate the crop's water use, the crop coefficient (Kc) is multiplied by the reference evapotranspiration (ETr) rate. This results in an accurate calculation of the crop water requirement for the specific day. To further refine irrigation recommendations, the application incorporates additional variables such as crop type, soil characteristics, farm location, and the type of irrigation system in use. These factors allow the app to provide precise irrigation requirements, including the amount of water needed and the optimal time for irrigation application. The application has been developed to cater to both English and Marathi-speaking users, ensuring accessibility for a broad range of farmers, progressive growers, and agricultural professionals. It has been distributed to farmers, scientists from State Agricultural Universities (SAUs), Krishi Vigyan Kendras (KVKs), non-governmental organizations (NGOs), and officers from the Department of Agriculture. By providing localized and accurate irrigation advice, this tool helps improve water use efficiency, optimize crop yields and promote sustainable agricultural practices across different regions.

GreenSeeker: The GreenSeeker is a compact, affordable optical sensor designed to assess and monitor the nitrogen requirements of crops through the use of NDVI (Normalized Difference Vegetation Index) readings. This lightweight and portable device is easy to use: it is positioned over the crop canopy, and once the trigger is activated, it captures precise NDVI measurements. The GreenSeeker works by emitting both red and infrared light, which interacts with the plant's surface. The device then measures the amount of each type of light reflected back from the crop. By analyzing this data, the GreenSeeker provides valuable insights into the health and nutritional status of the plants, specifically their nitrogen levels. With its ability to offer real-time, accurate readings, the GreenSeeker enables farmers to evaluate crop vitality and determine the precise amount of fertilizer needed for optimal growth. This targeted approach allows for more efficient fertilizer application, reducing waste and minimizing the environmental impact of excess inputs. In turn, this leads to improved crop yields, reduced input costs, and more sustainable farming practices. By giving farmers a clearer understanding of their crops' needs, the GreenSeeker helps to enhance overall crop management, making it a valuable tool for increasing productivity and promoting environmentally conscious agriculture.

SPAD meters: SPAD meters are portable, handheld diagnostic tools that are widely utilized in agriculture to measure the chlorophyll content in plant leaves without causing any damage to the plants. These lightweight and easy-to-use instruments offer an effective and non-destructive method of assessing the health of plants by providing a numerical value called the SPAD value. This value serves as a reliable indicator of the chlorophyll concentration in the leaves, which is essential for determining the plant's overall vitality and nutritional status.The working principle of the SPAD meter involves shining light through the leaf at two specific wavelengths: one in the red light spectrum and the other in the near-infrared spectrum. The meter then measures the amount of light absorbed by the leaf at these wavelengths, and this data is used to calculate the Chlorophyll Concentration Index (CCI). The ratio of the transmitted light at these wavelengths provides a precise estimation of the chlorophyll content in the leaf, which correlates directly to the plant's ability to perform photosynthesis. This method allows for the accurate and rapid assessment of chlorophyll levels in a non-invasive manner, making it a valuable tool for farmers, agronomists, and researchers. SPAD meters play an essential role in precision agriculture, helping to monitor plant health, detect nutrient deficiencies, and optimize fertilization practices. By providing real-time data on the chlorophyll content, these meters enable better decision-making in crop management, ensuring healthier plants, increased productivity and more sustainable farming practices. Their efficiency and ease of use make SPAD meters an indispensable instrument in modern agricultural practices.

. Specifically, the CCI is determined using the formula:

CCI = per cent transmittance at 931 nm / per cent transmittance at 653 nm

For a given crop species, a higher SPAD value typically indicates a healthier plant, as it reflects greater chlorophyll content. This makes SPAD meters invaluable for monitoring plant health, guiding nutrient management, and optimizing crop performance by providing immediate and actionable insights into chlorophyll levels.

Figure 7.

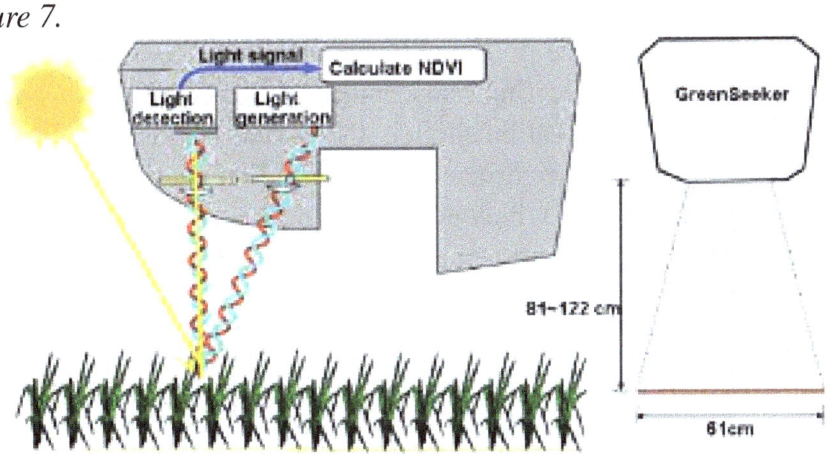

Figure 8.

Unmanned aerial vehicles: Unmanned Aerial Vehicles (UAVs), commonly known as drones, are remotely operated aircraft that can be controlled by a human operator or autonomously guided by an onboard computer system. These versatile devices are instrumental in collecting, assessing, and integrating a wide range of spatial, temporal, spectral, and continuous data. This capability is crucial for soil and water characterization as well as hydrological forecasting. The raw images captured by UAV sensors are typically processed to produce detailed orthomosaic images and surface models. Drones are highly valuable in managing water resources, as they enable precise mapping of wetlands, monitoring the extent of blue-green algae, and assessing water pollution levels or streamflows. Additionally, they are effective in detecting leaks within water distribution systems and updating digital elevation models that are essential for flood risk assessment. In agriculture, drones greatly simplify crop monitoring by gathering extensive data, which aids in accurate planning and decision-making. They facilitate improvements such as optimizing fertilizer applications, ensuring more efficient use of resources, and enhancing overall crop management practices. By integrating these advanced tools, drones significantly contribute to both water resource management and agricultural efficiency, making them indispensable in modern environmental and farming practices. The drone survey methodology includes:

- **Survey area selection and estimation**: The first step involves choosing the specific area for the survey and making preliminary estimates. This process ensures that the area selected aligns with the survey objectives and that the data collected will be relevant and comprehensive.
- **Flight planning**: Comprehensive flight planning is essential for the effective execution of the drone survey. This includes designing the flight path, setting the appropriate altitude, and adjusting camera settings to ensure thorough coverage and high-quality data capture.
- **Conducting the drone survey**: In this phase, the drone is flown over the designated area to capture high-resolution images and videos. This data collection is critical for detailed analysis and subsequent processing.
- **Processing images and creating orthomosaics**: After the survey, the captured images are processed to generate orthomosaics. This involves combining multiple images into a seamless, high-resolution map of the area and applying geo-referencing to align the map with precise geographical coordinates.
- **Overlaying village maps**: The orthomosaic is then superimposed onto existing village maps. This integration helps in aligning the new survey data with existing geographical information, improving the accuracy and usefulness of the results.
- **Assessing area and identifying crops**: The next step involves evaluating the area and identifying the different types of crops present. This assessment provides valuable insights into land use patterns and crop distribution.
- **Data entry into assessment form no. 2**: Finally, the analyzed data is recorded in Assessment Form No. 2. This form consolidates all the survey findings for documentation, reporting, and further analysis.

These steps ensure a thorough and accurate survey, facilitating better land and crop management through detailed and reliable data collection and analysis. This technique represents a significant advancement in agricultural technology due to its speed, accuracy, and minimal requirement for human labor. It operates with a high degree of transparency and is grounded in evidence-based practices. As a result, drone technology has proven to be highly effective for various applications including irrigation management, fertilizer spraying, crop monitoring, and aerial surveillance.

In a study carried out by Kolhe and Munde (2019), the effectiveness of using drone surveys for optimizing water management was thoroughly examined. The research specifically focused on evaluating irrigated regions and identifying the types of crops being cultivated within those areas. The study's findings demonstrated a significant increase in the total irrigated area, which grew from 10.71 lakh hectares in the previous year to 13.884 lakh hectares. This expansion of irrigated land translated into an additional assessment value of Rs 50 crores, highlighting the economic benefits of improved water management practices. The results of this study emphasize the powerful role that drone technology plays in enhancing the efficiency of water usage, enabling precise monitoring of agricultural land, and ultimately contributing to better crop management and overall agricultural productivity. The use of drones in such assessments has proven to be an effective tool for both improving resource management and maximizing the potential of agricultural land.

Heli-borne survey: Developed by the Council of Scientific and Industrial Research-National Geophysical Research Institute (CSIR-NGRI) in Hyderabad, this advanced geophysical mapping technique offers high-resolution 3D imaging of subsurface features up to 500 meters below the ground. The primary goal of this innovative technology is to map potential groundwater sources and improve their management, particularly to provide safe drinking water in the water-scarce, arid regions of India. This ambitious project, valued at Rs. 150 crores, is being rolled out in two phases. In its execution, CSIR has partnered with the Ministry of Jal Shakti under the "National Aquifer Mapping Project." The collaboration aims to enhance the identification and management of groundwater resources, thereby addressing critical water shortages in these vulnerable areas. By adopting this innovative technology, the Council of Scientific and Industrial Research (CSIR) will make a substantial contribution to the objectives of the Jal Jeevan Mission, which aims to enhance water accessibility and efficient management across India. This prominent initiative highlights a strong commitment to providing sustainable water solutions and emphasizes the importance of addressing water challenges at a national level. Moreover, it strategically positions CSIR as an essential contributor to the advancement of water resource management, further supporting India's long-term water security goals and ensuring the well-being of communities throughout the country.

Figure 9.

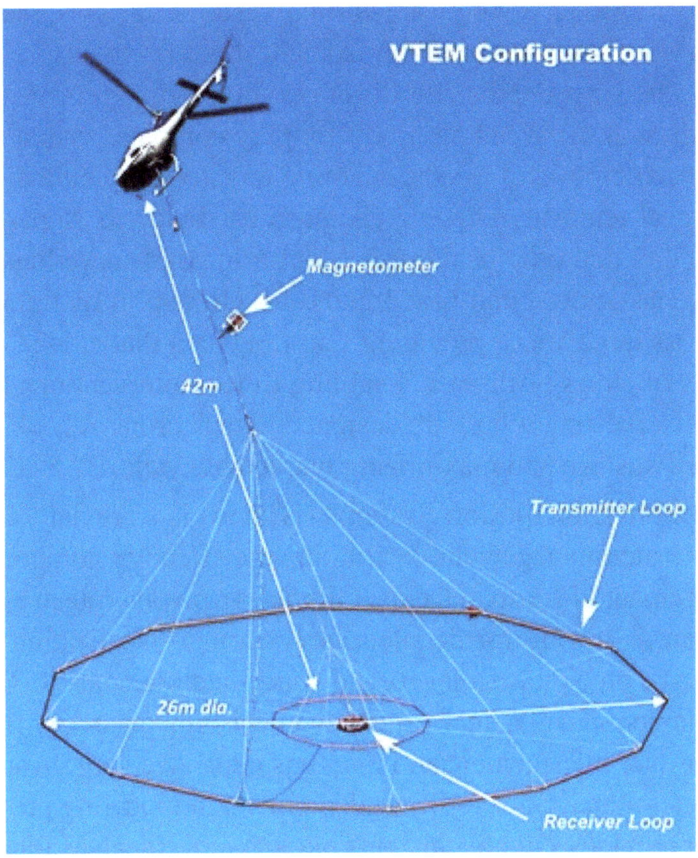

Agribots: An agribot are the "agricultural robot," that include autonomous tractors and drone monitors and do tasks like planting, weeding, monitoring and harvesting crop with precision and accuracy. These agribots play a pivotal role in reducing labour, enhancing efficiency and improving sustainability. is an automated solution designed to improve various farming operations. It offers precise seed placement at the optimal depths and spacing, enhancing germination rates and boosting crop yields. Equipped with advanced sensors and AI technology, these bots can differentiate between weeds and crops, allowing them to target weeds with pinpoint accuracy. This selective weeding approach reduces chemical use and promotes sustainable farming practices. Furthermore, agribots are adaptable to different crops and changing conditions, utilizing real-time data to perfectly time harvests, ensuring both the highest yield and the best quality of produce. There are various types of agribots like i) Planting robots which use advanced ultrasonic technology to ensure seeds are planted at the exact depths and distances, improving germination rates and crop growth. ii) Weeding robots employ computer vision to distinguish between crops and weeds, enabling targeted herbicide application or mechanical removal, reducing chemical use by up to 95 per cent and promoting soil and crop health. iii) Monitoring and surveillance robots include drones and ground-based units, which help in monitoring crop health across large areas, providing real-time data that aids in detecting early signs of stress or disease, enabling timely interventions to reduce losses. iv) Harvesting robots are designed to pick crops at peak ripeness, minimizing wastage and maximizing marketable yield.

Geographic Information Systems (GIS): It is a crucial tool in contemporary water management, facilitating the effective storage, analysis, and management of geospatial data. GIS plays a pivotal role in hydrologic management, allowing for detailed analysis and visualization of water flow patterns, precipitation levels and watershed boundaries. This helps water resource managers make informed decisions regarding flood control, stormwater management and river basin management. In addition to surface water management, GIS is also crucial in modeling groundwater systems. It provides valuable insights into the movement, availability and quality of groundwater, aiding in the assessment of aquifer levels, the detection of potential contamination risks and the sustainable management of underground water resources. Furthermore, GIS is instrumental in water quality analysis by integrating data related to pollutants, chemical composition, temperature and other environmental factors. This integration helps monitor and protect the health of water bodies, ensuring that they meet quality standards and are safe for consumption and ecological health.

Additionally, Geographic Information Systems (GIS) significantly enhance water supply management by optimizing the design and functioning of distribution networks, tracking water usage patterns, and predicting future water demands based on various factors such as population growth and climate change. By providing real-time data and advanced analytical tools, GIS plays a crucial role in improving the efficiency of water resource allocation, ensuring that water is distributed where it is needed most. Moreover, GIS supports long-term sustainability by identifying potential areas of concern and guiding proactive measures. Ultimately, the integration of GIS in water management leads to better-informed decision-making, the adoption of more sustainable practices, and enhanced protection and conservation of critical water resources for future generations.

Remote sensing: It has become a vital tool in water management, providing essential insights for the monitoring and optimization of water resources. One of its primary applications is in delineating surface water bodies, which includes the identification and accurate mapping of various water features such as lakes, rivers, reservoirs, ponds and wetlands. By using satellite imagery and advanced remote sensing techniques, water bodies can be tracked over time, providing crucial data for water resource allocation, flood risk assessment and ecosystem health monitoring. Additionally, remote sensing plays a significant role in estimating meteorological variables, which are essential for understanding local and regional climate conditions. By analyzing satellite-based weather data, remote sensing can provide accurate estimates of key meteorological factors like temperature, precipitation, wind speed and humidity. These variables are vital for water management strategies, as they directly influence the availability of water and the potential for drought or flooding. Another important application is the estimation of hydrological variables, which are critical for managing water systems. Remote sensing allows for the monitoring of soil moisture levels, streamflow and groundwater recharge, all of which are important for understanding the movement and availability of water across landscapes. This data aids in the development of more accurate hydrological models, improving decision-making for irrigation, flood control and watershed management. Furthermore, remote sensing technologies are highly effective in estimating evapotranspiration, the process by which water is transferred from the soil and vegetation to the atmosphere. By analyzing changes in land surface temperature, vegetation indices and atmospheric conditions, remote sensing provides reliable estimates of evapotranspiration rates. This information is essential for assessing crop water requirements, optimizing irrigation practices and monitoring water use efficiency in agriculture. It also helps in drought monitoring, as evapotranspiration data can indicate areas of water stress. Its ability to monitor and assess water-related parameters on a large scale and over time supports sustainable water resource management, agricultural planning and environmental protection.

3. ADVANTAGES AND DISADVANTAGES OF AI AND ICT

Advantages

- Improves the efficiency of decision-making by providing accurate and timely information, allowing people to make better and faster choices.
- Allows for real-time assessments, making it possible to analyze situations immediately and take quick action when needed.
- Offers clearer and more detailed insights, helping people understand situations better and make decisions based on solid information.
- Reduces environmental impacts by using resources like water, energy, and soil more efficiently, and by cutting down on waste.
- Lowers the need for manual labor by automating repetitive tasks, making operations smoother and reducing physical strain on workers.
- Saves time by automating processes and speeding up tasks, allowing work to get done faster and more efficiently.
- Ensures that important information is delivered quickly, improving how quickly people can respond to changes and make adjustments.
- Increases overall productivity, efficiency and sustainability for farmers, helping them achieve better results while managing resources more effectively and responsibly.

Disadvantages

- The high implementation costs associated with advanced technologies can pose a major challenge, necessitating a considerable financial investment for their acquisition, installation and seamless integration into existing systems, which may limit accessibility for some organizations.
- There is a shortage of experienced programmers who are skilled in developing and managing these complex systems, which can limit their effective utilization.
- The automation and efficiency gains from these technologies can lead to job displacement, potentially causing unemployment as roles traditionally performed by humans are replaced by machines.
- The technology can be challenging to understand and operate, requiring specialized knowledge and training to use effectively.
- If these technologies fall into the wrong hands or are misused, they have the potential to cause significant harm or disruption.
- The systems are limited to performing tasks within the scope of their programming and cannot adapt or function beyond the specific parameters set by their developers.

CONCLUSION

In the contemporary digital era, the integration of cutting-edge technologies with progressive management practices, policy reforms, and institutional changes has the potential to revolutionize agricultural practices and significantly enhance the livelihoods of farmers. By leveraging the power of pioneering artificial intelligence (AI) techniques and advanced information and communication technologies (ICT), the agriculture sector can achieve unprecedented efficiencies and advancements. These technological innovations are transforming the way farming is done, providing valuable tools that improve productivity, reduce costs, and minimize the wastage of critical resources.AI and ICT technologies offer a wide range of applications that streamline numerous aspects of farming, leading to optimized resource management. With the capability to process vast amounts of data in real time, these tools enable farmers, agricultural scientists, and extension workers to make more informed decisions about crop management, irrigation, fertilization, and pest control. The real-time data analysis offered by AI and ICT technologies ensures that decisions are precise and timely, leading to better management of land and water resources. By reducing inefficiencies and wastage, these technologies play a pivotal role in improving agricultural outcomes while minimizing environmental impact.

This technological shift represents a profound change in agricultural practices, fostering a more targeted and effective approach to farming. AI and ICT enable agricultural professionals to interact with the farming community in a more personalized and informed way, providing farmers with tailored advice and data-driven insights that optimize crop yields and resource utilization. The ability to predict crop performance and monitor environmental conditions in real time has made it easier to enhance irrigation schedules, optimize nutrient application, and anticipate potential yield outcomes, all of which contribute to more efficient and sustainable farming practices.By enabling smarter decision-making and facilitating better management of resources, AI and ICT technologies not only enhance productivity but also contribute to building a more resilient agricultural sector. These advancements support broader goals such as ensuring food security, increasing economic growth, and enhancing sustainability. The widespread adoption of these technologies marks a significant evolution in agricultural methodologies, offering new opportunities for progress. As agriculture continues to embrace the digital revolution, these innovations are poised to reshape farming, creating a more efficient, sustainable, and prosperous agricultural sector capable of meeting the challenges of a rapidly changing world. The potential of AI and ICT to drive transformation in agriculture is immense, offering a future where farming is not only more productive but also more environmentally responsible.

REFERENCES

Anonymous. (2019). *Potential of ICT in water sector*. https://www.teriin.org/events/icdl/sdg-files/pdf/ICT-in-water-sector.pdf

Barkunan, S. R., Bhanumathi, V., & Sethuram, J. (2019). Smart sensor for automatic drip irrigation system for paddy cultivation. *Computers & Electrical Engineering*, *73*, 180–193. DOI: 10.1016/j.compeleceng.2018.11.013

Das, S., Munshi, M. N., & Kabir, W. (2017). The impact of ICTs on agricultural production in Bangladesh: A study with food crops. *SAARC Journal of Agriculture*, *14*(2), 78–89. DOI: 10.3329/sja.v14i2.31247

Doorn, N. (2021). Artificial intelligence in the water domain: Opportunities for responsible use. *The Science of the Total Environment*, *755*, 1–10. DOI: 10.1016/j.scitotenv.2020.142561 PMID: 33039891

Food and Agriculture Organization of the United Nations. (2019). *Food and Agriculture Organization of the United Nations*. https://www.fao.org/fao-stories/article/en/c/1185405

Kamil, I. A., & Ojoawo, O. J. (2019). Smart irrigation: An ICT application in agriculture. *The International Journal of Engineering and Science*, *8*, 53–58.

Kolhe, P., & Munde, T. N. (2019). Use of drone for efficient water management: A case study. *Proceedings of the 3rd World Irrigation Forum*, 1–13.

Liu, Z., & Xu, Q. (2018). An automatic irrigation control system for soilless culture of lettuce. *Water (Basel)*, *10*(11), 1692. DOI: 10.3390/w10111692

Malviya, A., & Jaspal, D. (2021). Artificial intelligence as an upcoming technology in wastewater treatment: A comprehensive review. *Environmental Technology Reviews*, *10*(1), 177–187. DOI: 10.1080/21622515.2021.1913242

Miller, L., Vellidis, G., Mohawesh, O., & Coolong, T. (2018). Comparing a smartphone irrigation scheduling application with water balance and soil moisture-based irrigation methods: Part I—Plasticulture-grown tomato. *HortTechnology*, *28*(3), 354–361. DOI: 10.21273/HORTTECH04010-18

Mohammad, F. S., Al-Ghobari, H. M., & El Marazky, M. S. A. (2013). Adoption of an intelligent irrigation scheduling technique and its effect on water use efficiency for tomato crops in arid regions. *Australian Journal of Crop Science*, *7*(3), 305–313.

Navarro-Hellin, H., Rincon, M. D., Martinez-del-Rincon, J., Sotos, V. F., & Torres, S. R. (2016). A decision support system for managing irrigation in agriculture. *Computers and Electronics in Agriculture*, *124*, 121–131. DOI: 10.1016/j.compag.2016.04.003

Saab, M. T., Jomma, I., Skaf, S., & Fahed, S. (2019). Assessment of a smartphone application for real-time irrigation scheduling in Mediterranean environments. *Water (Basel)*, *11*(2), 2–18.

Singh, B., Singh, V., Purba, J., Sharma, R. K., Jat, M. L., Singh, Y., Thind, H. S., Gupta, R. K., Chaudhary, O. P., Chandna, P., Khurana, H. S., Kumar, A., Singh, J., Uppal, H. S., Vashistha, M., & Gupta, R. (2015). Site-specific fertilizer nitrogen management in irrigated transplanted rice (*Oryza sativa*) using an optical sensor. *Precision Agriculture*, *16*(4), 455–475. DOI: 10.1007/s11119-015-9389-6

Singh, J., Heeren, D. M., Ge, Y., Geng, B., Neale, C. M. U., Maguire, M. S., & Bhatti, S. (2021). Sensor-based irrigation of maize and soybean in east-central Nebraska under a sub-humid climate. *Biological Systems Engineering: Papers and Publications*, *763*, 1–12. DOI: 10.13031/aim.202101044

Singh, K. M., Kumar, A., & Singh, R. K. P. (2017). Role of information and communication technologies in Indian agriculture: An overview. SSRN *Electronic Journal*, 1–12.

Sinwar, D., Dhaka, V. S., Sharma, M. K., & Rani, G. (2019). AI-based yield prediction and smart irrigation. *Springer Nature*, *2*, 155–179.

Zia, H., Rehman, A., Harris, N. R., Fatima, S., & Khurram, M. (2021). An experimental comparison of IoT-based and traditional irrigation scheduling on a flood-irrigated subtropical lemon farm. *Sensors (Basel)*, *21*(4175), 3–17. DOI: 10.3390/s21124175 PMID: 34204584

Chapter 3
A Review on Artificial Intelligence for Smart Agriculture

N. Suthanthira Vanitha
https://orcid.org/0000-0002-9579-7651
Solamalai College of Engineering, India

K. Radhika
https://orcid.org/0000-0003-4849-8680
Muthayammal Engineering College, India

J. Gowrishankar
https://orcid.org/0000-0002-4320-8683
Jain University, India

C. S. Satheesh
Muthayammal Engineering College, India

G. Sudha
https://orcid.org/0000-0002-1434-9188
Muthayammal Engineering College, India

M. Maheshwari
University of Technology and Applied Sciences, Oman

ABSTRACT

The development of Artificial Intelligence Techniques has created modernization in the agricultural domain. The agricultural sector needs innovation to survive with the varying climatic conditions and other environmental factors. The factors like, Soil Prediction, Plant Condition, Weather and Crop Yield plays a vital role in smart farming. Uncertainty in the above parameters makes agriculture very critical. Machine Learning Techniques

DOI: 10.4018/979-8-3693-4392-0.ch003

support conventional farming in a cost-effective approach. The implementation of Machine Learning in the agricultural aspects is to increase crop growth, create a subtle environment for the crop, change plant cultivation to produce more with less resources and crop cultivation management process. In addition, Deep Learning algorithms in agricultural field provide better results in crop production and productivity. This chapter reviews about role of Artificial Intelligence and applications in Smart Agriculture.

1. INTRODUCTION

In the global economy, agriculture plays a vital role to implement scientific techniques and in various aspects of research. Due to population growth, there is a continuous demand in the agricultural sector to enhance the production of more crops. Climatic change is caused as a result of growth in population, pollution, solid wastage, decrease in groundwater level as well as the global environment (Thayer et al, 2020; Nassani et al, 2019). The environmental factor is one of the major obstacles which affect agricultural production. The factors are weather conditions, intra-seasonal variability, and spatial variability of soil properties. Recent innovations in research support carrying out crop yield development, climatic changes, crop prediction, compost identification, weed harvest, and insect detection (Conrad et al, 2018).

The agriculture industry is relatively quite slow for farmers which make them unable to invest in emerging tools and technologies. Digital Agriculture in farming uses a data-intense approach that drives agricultural productivity by minimizing environmental impact (Benos et al, 2020).In the case of agricultural operational environments, Machine learning with data analytics and high-performance computation generates new prospects to unravel, measure, and realize data-intensive processes (Sorensen et al, 2019).The performance evaluation of predictive models is determined by the training and testing phase. Yield prediction acts as a tool for the optimization and evaluation of crop-area insurance contracts.

The contribution of computer science in agriculture has created a significant impact on artificial intelligence, machine learning, deep learning, quantum computing, data analytics, image processing, and computer vision

(Sonka, 2016). To forecast crop growth, disease detection, and fertilizer for specific plants Machine Learning techniques support farmers in the agriculture field. In addition, to harvest yield prediction, crop quality estimation, categorization of plant species, crop disease detection, and weed infestations ML-enabled sensors are used. Generally in yield prediction, datasets like patterns, correlations, and knowledge discovery are used to train the features which are compared with the outcomes of experience (Helm et al, 2020).

2. LIFE PROCESS IN AGRICULTURE

Figure 1 shows the life process in agriculture.

It includes the following process.

Land Plowing: The primary process of farming is to prepare the soil for seed sowing which involves flouting soil clumps, and removing garbage like sticks, rocks, and roots. For increasing the strength of the soil the farmers add organic manure and fertilizer depending on the type of crops.

Seed Sowing: This process requires the quantity, depth, and distance of planting seeds. The effective growth of the seed is based on climatic conditions such as temperature, humidity, and rainfall.

Figure 1. Life process in agriculture

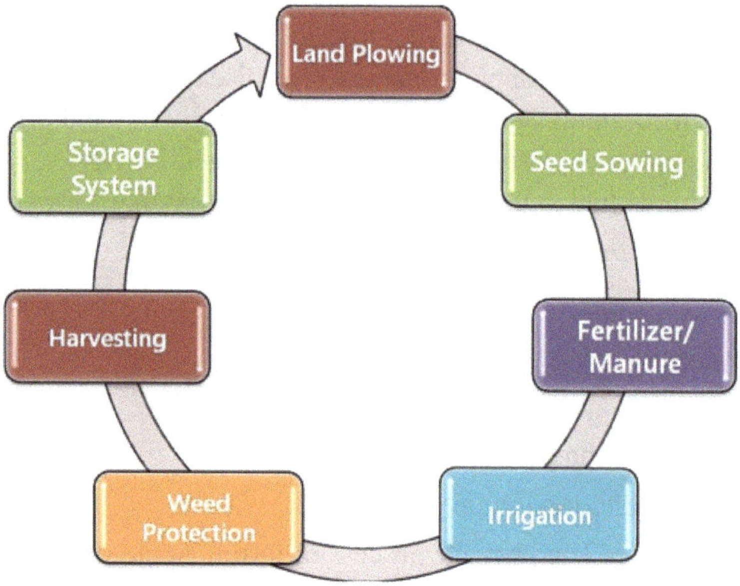

Fertilizer/Manure: To maintain the fertility of the soil, manure is used for cultivation. In general, farmers use fertilizers to increase the nutrients of the plant. The nutrients such as nitrogen, phosphorus, and potassium with their high potential help large crop production.

Irrigation: This process supports maintaining soil moisture and humidity. Underwatering can slow down the growth of crops and if the irrigation is not proper that leads to a reduction in crop yield.

Weed protection: In this process, weed protection is considered an important factor for plant growth. Weed decreases plant growth in the boundary of farms, high production costs, harvest interference, and lesser crop quality.

Harvesting is the process of assembling the field crops after they mature. As part of the labor-intensive activity, it involves more number laborers for harvesting. And it has a post-harvest handling process like cleaning, sorting, packing, cooling, etc., depending on the type of crop.

Storage System: This stage includes the post-harvest method in which harvested crops are kept in the storage system. It can be maintained with proper precautions and also it covers packaging and crop transportation

From the above process, it is clear that the conventional agricultural domain has challenges due to changes in climatic conditions, deficiency of nutrients for specific crops, and weed protection.

3. MACHINE LEARNING IN AGRICULTURE

3.1 Machine Learning Terminologies

ML methodologies use a set of data to carry out a task. The information in ML has a collection of example data. Normally an entity of data is explained by a set of variables or attributes. Attributes can be categorized as nominal, binary, and numeric values. The specific task of the ML model is calculated based on the performance metrics which are enhanced with experience over time. ML algorithms are estimated with the support of statistical and mathematical models. Figure 2 depicts the machine learning system's general initiatives. By pre-processing, the raw input data is derived into a proper state, which includes data cleaning by removing inconsistent data, integration of data, and data conversion (Anagnostis et al, 2020). For the implementation of the learning model, the feature extraction and selection create the informative subset of features (Zheng et al, 2018). Overall learning performance is improved by the feedback system. The output is obtained by applying the unobserved testing data with the feature vectors of the trained model, through the prediction rule.

Figure 2. General machine learning system (Vanitha et al, 2023)

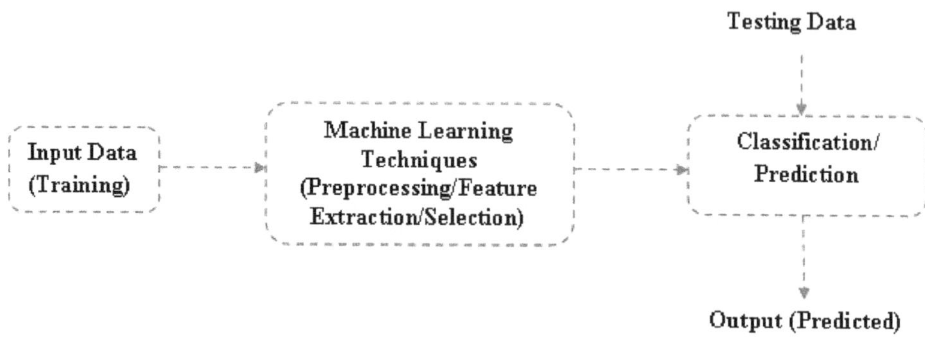

Machine learning along with data analytics is growing to create new prospects to realize various environmental data processes in agriculture. Machine learning technology has the ability that permits machines to learn without programming the devices. To manage the Modern agriculture challenges such as food scarcity, population growth, safety and health hygiene, natural resources reduction, dietary adaptation, and weather changes. To overcome the above issues, the agricultural sector is forced to focus on optimizing the efficacy of agricultural practices which reduces the troubles in the environment. In specific, these two prerequisites impel the revolution of agriculture into smart agriculture. This transformation assures sustainability, maximum production, and a safe environment in farming (Lampridi et al, 2019).

Generally, smart farming has four parameters as shown in figure 3 to compete for the growing needs (Zecca, 2019).

(i) Natural Resource Management

(ii) Development Services
(iii) Ecosystem Conservation
(iv) Modern Technologies

Figure 3. Smart farming parameters (Zecca, 2019)

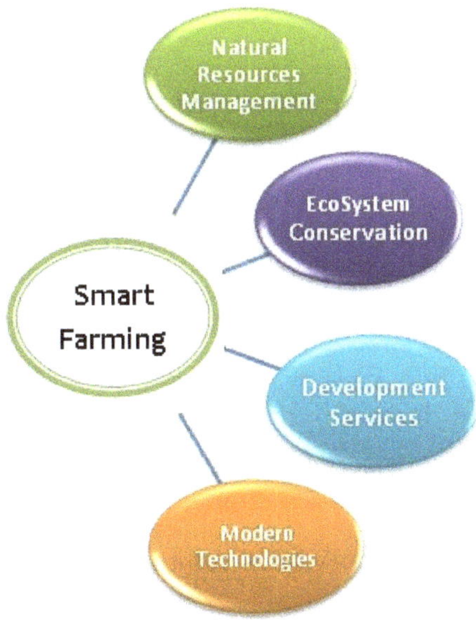

The adoption of ICT in modern crop growing includes sensors, cameras, satellites, wireless sensor networks, drones, autonomous vehicles, etc., for safety, security, and economically feasible (Sorensen et al, 2019). Big data is meant for a huge quantity of data formed by digital technologies. In addition to editing, analyzing and interpreting big data requires huge storage capacity. Interpretation in big data has sizeable potential to add value for society and the environment (Sonka, 2016). However, big data has certain challenges while extracting the data which include Variety, Velocity, Volume, Veracity, and Value (Liakos et al, 2018).The traditional data processing techniques remain ineffective to meet the growing demands of smart farming, it acts considerable difficulty in extracting important information from the agricultural field. Machine learning is used for the growth of exponential computational power capacity (Helm et al, 2020). There are plenty of applications of ML in agriculture.

3.2 Machine Learning Algorithms

ML is the domain of artificial intelligence that provides information to machines and can learn statistical data instead of program coding (Samuel, 1959).ML is categorized into three types namely, supervised learning, unsupervised learning, and reinforcement learning shown in figure 4. Supervised learning is referred to as; the single supervisor used to train the data through training examples. It supports labeling datasets for algorithm training, data classification, and data prediction accurately. Examples of Supervised Learning are Artificial Neural networks, Bayesian Networks, Decision Trees, Support Vector Machines (SVM), ID3, k-Nearest Neighbor, Markov Model, etc. Unsupervised Learning is referred to as an infinite quantity of data to the program for pattern findings and patterns relationship. By using unsupervised learning hidden patterns in the datasets can be discovered. For example, k-Nearest Neighbor, k-Means Clustering, etc.

Reinforcement learning is one of the methods of supervised and unsupervised learning that creates a sequence of decisions by training machine learning models. Reinforcement learning-based robots are used in the industry for performing several applications. With the rise of agricultural data, to obtain precise crop prediction in agricultural production, Machine learning techniques are used. The following figure 4 depicts the machine learning algorithms.

Figure 4. Classifications of Machine Learning Algorithms (Suthanthira Vanitha et al, 2023).

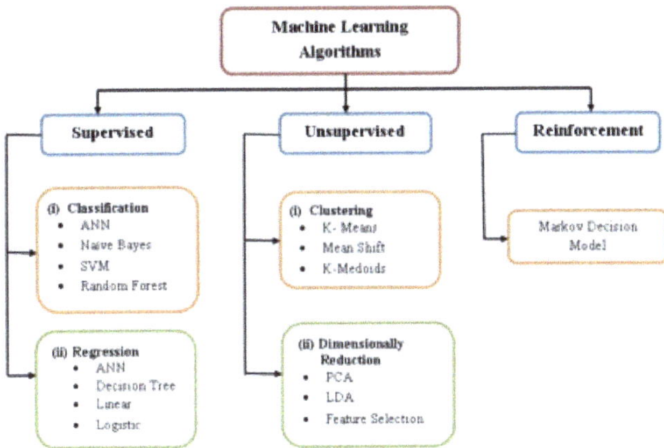

1. Artificial Neural Networks

ANN is similar to the human brain, formed by biological neurons based on various structures and functionalities. Due to its well-organized structure, it is used as a data-driven modeling tool for dynamic modeling and the identification of nonlinear systems. For example, significant results are created for the incomplete data, the data is trained using neural networks for predicting future data with similar patterns. Precision can be increased by adding additional data. Without knowing the basic principles ANN can adopt complex data. In any process, it can capable of developing the relationship between input and output data. Monisha Kaul et al. (2005) applied an artificial neural network a model has been developed for corn and soybean yield based on climate conditions. By considering the rainfall, corn and soybean yield can be calculated at various levels with the support of ANN and multiple regressions. Finally, while comparing, ANN performance provides better results for yield than regression techniques.

Dahikar & Rode (2014) used artificial neural networks for crop prediction through soil and climate parameters. It includes the soil nutrients, temperature, rainfall, humidity, etc. Testing is carried out on the crops such as Corn, Rice, Cotton, Sugarcane, Soybean, Wheat, and Groundnut. In Stathakis et al. (1994), the prediction of crop yield parameters is analyzed through remote sensing neural networks that can be applied to the Neuro-Fuzzy Inference System (ANFIS). ANFIS input parameters like soil moisture, ground temperature, and repository organs are obtained only a single output (crop yield). The efficiency of the crop yield is obtained using ANFIS. Crop prediction can be influenced by various factors. Uno (2005) used artificial neural networks and decision trees to develop yield mapping and forecasting for determining the accuracy of agronomic variables, nitrogen application, and weed control.

2. Decision Trees

Includes various parameters like nodes, branches, and terminal values. Here the datasets are organized in minor homogeneous subsets, while an allied tree diagram is produced at the same time. The internal node corresponds to a selected feature by comparing with different pairs of data, while the branch stands for the output in the regression model. Based on the classification rule the Leaf node signifies the prediction of the data. The process has two steps for the construction, the initial phase - is large growth then size reduction, and over data fitting, whereas the next phase is used for classification purposes described in Vijaysingh & Solanki (2014).

Veenadhari et al. (2011) used the decision tree algorithms for soybean productivity modeling. Considering weather factors such as evaporation, temperature, and rainfall. The Interactive Dichotomizer3 (ID3) algorithm has two characteristics. The major influencing parameter on the soybean crop yield as per the induction tree analysis report is moisture. A decision tree is formulated based on the classification of if – then – else rules. The obtained results show that the production of soybean is affected by relative humidity and that can be overcome by generating some rules which support the prediction of low and high soybean yield. But the total production of soybean yield is unpredictable.

3 . Support Vector Machines

SVM is meant for classifying the data in binary form that makes a linear separating hyperplane. By using the kernel trick, the conventional SVM classification significantly improved the dimension of an original feature. Normally, it is classified as classification, regression, and clustering. support vector regression (Smola, 1999), least squares support vector machine (Suykens et al, 2002), and successive projection algorithm-support vector machine (Galvao et al, 2008) are common SVM algorithms. Smart algorithms are employed in precision agriculture to meet the present agriculture requirements. The SVM classifier is used in sugarcane farming to increase productivity, safe and secure environment for continuous monitoring of parameters, infection identification, and healthy crop growth .

4. Bayesian Network

Bayesian network (or) Bayesian model is the class of supervised learning based on a statistical analysis of the attribute in an assigned dataset. The probabilistic-directed acrylic model provides the solution for the classification and regression problems. Some of its well-known algorithms are Naive Bayes, Gaussian naive Bayes, multinomial naive Bayes, Bayesian network, and the mixture of Gaussians and Bayesian belief networks. In this system, the data is characterized in charts that are directed by nodes in which the nodes are meant for function whereas the edges are on data dependency. Yiqun Gu et al. (1994) used Bayesian model is to evaluate the effect of climatic change on potato production. By taking the inconsistency of present climate parameters such as rainfall, temperature, radiation, and the awareness of potato growth. The test results conclude that the efficiency is more when compare conventional mathematical model.

5. Random Forests Algorithm

Random forests algorithm is used for classification and regression as that of supervised learning. A group of learning methods that consists of several decision trees improves the prediction by averaging the impact of many decision trees. Other models are also available for predictions on data classification. Random forest classification algorithm produces a high accuracy with less processing time for agricultural data analysis.

6 . K- Means Clustering

In Shaikh & Mukane (2017), this method is applied to the pomegranate plant disease for image segmentation. The collection of healthy and unhealthy plant images is the dataset. A diseased leaf is identified by preprocessing and segmenting the data into three sub-images that contain dissimilar features. Consequently, this method of detecting does not give a large accuracy.

7. Markov Chain Model

Markov Chain Model is a type of reinforcement learning in machine learning techniques. It is a collection of a random variable process that contains the analytical model in a probabilistic manner in which output depends on the previous results. Matis et al. (1989) applied Markov chain approach for the forecasting of cotton yield with multiple linear and rank regressions models. Forecasting of sugarcane yield second-order Markov Chain Model is applied in which the data are used simultaneously. The result of crop yield forecasting concludes that the efficiency obtained from the second-order Markov Chain Model is better than the first-order.

4. DEEP LEARNING IN AGRICULTURE

Recent advances in image processing and data analysis have given rise to deep learning, a technique with great potential and promising outcomes. Recently, deep learning has made its way into the agricultural industry, following its successful application in other fields. Feature learning, or the automatic extraction of features from raw data, is a powerful benefit of deep learning (DL). Features from lower levels of the hierarchy are produced by the composition of features from lower levels (Senthil Kumar et al, 2022). Since more complicated models are used, which enable huge parallelization, DL is capable of solving more complex problems very well and promptly (LeCun et al,2015). If sufficiently extensive datasets characterizing the problem are available, these sophisticated models used in DL can improve classification accuracy or decrease error in regression situations.

Depending on the network architecture being utilized (i.e., Unsupervised Pre-trained Networks, Convolutional Neural Networks, Recurrent Neural Networks, Recursive Neural Networks), DL consists of a variety of different components.Due to their extensive learning capacity and highly organized structure, deep learning models are particularly good at classification and prediction, making them versatile and adaptable to a wide range of extremely complicated problems from the perspective of data analysis (LeCun et al,2015). Even though DL has gained popularity in many applications involving raster-based data (such as images and video), it can also be used with any type of data, including speech, natural language, audio, and more broadly, continuous or point data like population, weather, and soil chemistry data and more (Pan & Yang, 2010; Sehgal et al., 2017;Demmers et al,2012).

5. APPLICATIONS OF MACHINE LEARNING IN AGRICULTURE

According to Liakos et al. (2018), four general categories utilize machine learning techniques as shown in figure 5. Types are as follows.

(i) Crop Management
(ii) Water Management
(iii) Soil Management
(iv) Livestock Management

Figure 5. Generic categories of agriculture (Liakos et al, 2018)

5.1 Crop Management

The objective of crop management is to achieve quantitative and qualitative targets (Yvoz et al. 2020). It involves the combination of farming methods in biological, chemical, and physical crop environments. The latest approaches to supervising crops such as crop prediction, disease, and weed detection, crop identification, and crop features increase farming production accordingly the financial income. The above features constitute the key objective of agriculture.

(i) Yield Prediction

Yield prediction is considered an essential and challenging issue in the agricultural field. The precision model supports the farmers in take decisions for choosing the crop to meet the current market demands (Van Klompenburg et al. 2020). Yield prediction accuracy is obtained by various factors such as environment, phenotypic characteristics, crop genotypic, and management practices. Therefore, it requires a basic relationship between the factors and the crop yield. Machine Learning techniques are applied to identify the type of relationships that mandates comprehensive datasets along with influential algorithms (Khaki & Wang, 2019).

(ii) Disease Detection

In agricultural production systems crop disease place a vital threat that depreciates the crop quality, and yield quantity at the storage and transportation stage. In the farming stage due to the various plant diseases, the farmers are facing yield losses (Harvey et al. 2014). Globally farmers pose a significant risk to food security due to crop diseases. The key aspect of efficient management is to identify the plant disease at an earlier stage. Plant diseases are aggravated by different types of bacteria, fungi, pests, viruses, and other agents. The plant disease symptoms are recognized through the presence of pathogens, changes in the plant phenotype, leaf and fruit spots, wilting, color change leaf curling (Jim Isleib, 2021). Generally, plant disease is detected by agronomists through field inspection for the time consuming of the crop. With the advancement of sensing systems, that identifies the plant disease before the symptoms are exposed to the environment. In recent years, Machine Learning techniques are employed which made extraordinary development in disease detection. For large-scale farming, autonomous vehicles are employed with Machine Learning algorithms for the timely identification of phytopathological harms (Anagnostis et al,2021).

(iii) Weed Detection

Normally weeds grow and spread out faster over larger areas of the field. It competes with crop resources like sunlight, space, water, and nutrients. Usually, weeds begin to grow frequently which harmfully affects crop growth (Su, 2020). To increase crop production weed control management is performed by either mechanical or herbicides. Mechanical management is critical and unsuccessful unless performed properly whereas herbicides are used widely in the weed treatment process. While using herbicides for a larger amount seems to be costlier and unfavorable for the atmosphere but for similar applications not including the spatial distribution of the weeds. However, the use of long-term herbicides makes weeds more resistant which results in demand and expensive weed control. With innovation in smart culture, substantial achievements pretend to differentiate weeds from crop yield (Islam et al., 2021). ML techniques are used for real-time differentiation, and localization of weeds, facilitating the accurate application of herbicides to a particular region, instead of targeting the complete field (Slaughter et al., 2008).

(iv) Crop Recognition

Scientific areas such as botanical gardens, plant taxonomy, and new species discovery have expanded sizeable interest in the automatic recognition of crops. The analysis of a mixture of organs including leaves, stems, fruits, flowers, roots, and seeds can be recognized and classified in plant species (Bonnet et al., 2016; Seeland et al., 2017). Leaf-based plant recognition is a common approach by examining the characteristics of leaves like color, shape, and texture in specific (Zhang et al., 2020). The advancement of machine learning techniques leads to the automatic recognition of crops from remote locations.

(v) Crop Quality

Generally crop quality depends on the soil and climate conditions, nature of cultivation, and characteristics of the crop. Farmers earn larger with high-quality agriculture products. Harvesting of the crop is based on the fruit quality, firmness, soluble solid content, and skin color (Papageorgiou et al., 2018). Both high-value and arable crops are affected by the time constraints of the harvested product characteristics. To overcome the above issue, the farmers need decision support systems for suitable management decisions for increasing the quality of crop production.ML algorithms with image processing techniques provide the solution for improving crop quality.

5.2 Water Management

Water management is a challenging factor in agriculture. Plant growth mainly depends on the availability of water resources therefore efficient water management is desirable for sustainable crop production (Neupane & Guo, 2019). In addition, successful water management improves water quality and reduces pollution and health threats (El Bilali et al., 2021). Recent research says, for accomplishing water savings, variable rate irrigation is projected in smart agriculture. This irrigation can be employed at various rates according to the water necessities of management zones. The success and viability of this type of irrigation depend on agronomic factors, which include the property of soil, topography, and soil water to achieve yield optimization and saving of water (Gunasekaran et al, 2024). Temporal and spatial patterns, condition of crop growth, soil water status monitoring, weather conditions, and forecasting assist irrigation for resourceful water management. For obtaining accurate water management, ICT offers spatial and temporal changeability images linked with the soil moisture status and crop growth parameters.

5.3 Soil Management

Soil management plays an important factor in the soil strength monitoring system. The type of soil and nutrition of soil is considered a significant feature in crop quality. With the escalation of deforestation, soil quality is degraded and it's tough to determine the soil quality. Soil management has complex mechanisms and processes due to its heterogeneous nature. For sustainable agriculture, soil management provides specific information with consistent land potential (Lampridi et al., 2019). Conventional soil evaluation techniques comprise oil sampling and laboratory analysis which offers economic remote sensing systems and soil mapping sensors for the easy result. ML techniques recognize soil nutrient deficiency, plant pests, and plant diseases through which farmers choose the fertilizers that support improving the harvest quality.

5.4 Livestock Management

Livestock Management influences consumer insight into food safety, security, and sustainability that mainly depends on animal welfare and human health and also cover social concern. Livestock production is widely acknowledged based on the strength of productivity per animal. In specific monitoring, the welfare of animals and net production is considered a key aspect to progress production systems. The main aim of precision livestock farming is to apply real-time engineering techniques for monitoring animal health, and message alerts and to increase the production rate at the earlier stage. Further precision livestock farming supports the farmers in the decision-making process. Non-invasive sensors are used in Precision livestock farming (Li et al., 2020; Gunasekaran et al., 2024). Variable physical quantities such as temperature, sound, humidity, etc are measured by IoT sensors. This sensor can inform whenever the physical quantities are changed beyond normal limits in real-time for the individual animal which reduces the difficulty of examining every animal (Akhigbe et al., 2021).]. Hence, to handle huge data ML methods have to turn out to be a fundamental part of modern livestock farming.ML algorithms support to definition of the behavior of biological system prediction.

6. CHALLENGES IN SMART AGRICULTURE

With an increase in population, there is a need to increase the agricultural system. Machine learning in agriculture is one of the fastest emerging fields. To improve the accuracy and to provide the solution for current problems ML techniques are used in the agricultural sector. Agriculture is a challenging field that comprises numerical data quantification. Environments are continuously changing from one division to the subsequent for a single field. The weather, soil quality changes, diseases, and pest possibilities are unpredictable. Farmers expect prospects are good for a forthcoming harvest, but the outcome is uncertain.

Naturally, agriculture arises among ecosystems of interacting organisms, crop production, and the nature of the activity that appears in the environment. However, these are not limited to weather manifestations. Therefore, managing an agricultural environment has many challenges. The factors that impact variance include rainfall measurement, soil nature, soil degradation pattern, and temperature. Hence, the issue with machine learning in agriculture is not because researchers have less potential for coding and procedures to deal with the large farmer problems, testing, validation, and technologies.

Machine Learning is to rectify the issues related to the physical atmosphere of all aspects of the physical interaction. Machine Learning teaches us many things to understand our environment, and to calculate significant yields in the agriculture field through the machine's cognitive response. Imagery-based Insights technology is an innovation in showing immense promise. In solving social, economic, climate change, and land use challenges, data collection bandwidth plays a part in agriculture farming.

7. CONCLUSION

The opportunities for Artificial Intelligence in the agriculture field are endless. The influence of ML in the development of farming supports the identification of the significant areas needed for technological innovation. Machine learning in agriculture not only supports farmers in automatic farming but modify accurate cultivation. With limited resources, higher-yield quality crops are produced. In addition, Agricultural Industries started focusing on machine learning-based products or services like training data for agriculture, drones, and autonomous vehicles. The integration of AI in agriculture supports forecasting crops by applying a suitable machine learning algorithm that contains certain merits over the conventional forecasting method.

ML and DL techniques are applied in several farming applications like crop price, crop management, crop yield prediction, Weed detection, disease identification, and other prediction like rainfall, temperature, moisture, etc. It supports high recommendations to farmers for proper decision-making. In addition, ML techniques have made predictions easier and government agencies to increase crop production. Data analysis offers emerging concepts of numerous crop management applications. Hence a majority of the research prefers ANN and SVM algorithms. In particular, ANNs are used in the crop, water, and soil resource management, whereas SVM is for livestock management. With the use of ML sensing data, farming methods are developing into actual artificial intelligence, that offers better suggestions and looming decisions for future crop production.

In the future, ML models are expected to be general and have the opportunity for integration of application tools. The combination of ML implementation, automation in data recording and analysis, and decision-making guide practical tolls streak through knowledge-based agriculture for growing quality bio-products. Due to advancements in agricultural technology, this sector supports the world to deal with food production issues for the growing population. In recent days there is rising attention to pests, weed detection, and weed sprayers in farming using computer vision technology in agricultural operations reduces stress levels. This chapter focused on Machine Learning applications in Agriculture, Algorithms, Deep learning, upcoming challenges and future perspectives in the agricultural field.

REFERENCES

Akhigbe, B., Munir, K., Akinade, O., Akanbi, L., & Oyedele, L. (2021). IoT Technologies for Livestock Management: A Review of Present Status, Opportunities, and Future Trends. *Big Data and Cognitive Computing*, *5*(1), 10. DOI: 10.3390/bdcc5010010

Anagnostis, A., Papageorgiou, E., & Bochtis, D. (2020). *Application* of Artificial Neural Networks for Natural Gas Consumption Forecasting. *Sustainability (Basel)*, *12*(16), 6409. DOI: 10.3390/su12166409

Anagnostis, A., Tagarakis, A. C., Asiminari, G., Papageorgiou, E., Kateris, D., Moshou, D., & Bochtis, D. (2021). A deep learning approach for anthracnose infected trees classification in walnut orchards. *Computers and Electronics in Agriculture*, *182*, 105998. DOI: 10.1016/j.compag.2021.105998

Benos, L., Bechar, A., & Bochtis, D. (2020). Safety, and ergonomics in human-robot interactive agricultural operations. *Biosystems Engineering*, *200*, 55–72. DOI: 10.1016/j.biosystemseng.2020.09.009

Bonnet, P., Joly, A., Goëau, H., Champ, J., Vignau, C., Molino, J. F., Barthélémy, D., & Boujemaa, N. (2016). Plant identification: Man vs. machine. *Multimedia Tools and Applications*, *75*(3), 1647–1665. DOI: 10.1007/s11042-015-2607-4

Conrad, Z., Niles, M. T., Neher, D. A., Roy, E. D., Tichenor, N. E., & Jahns, L. (2018). Relationship between food waste, diet quality, and environmental sustainability. *PLoS One*, *13*(4), e0195405. DOI: 10.1371/journal.pone.0195405 PMID: 29668732

Dahikar, M. S. S., & Rode, S. V. (2014). Agricultural crop yield prediction using an artificial neural network approach. *International Journal of Innovative Research in Electrical, Electronics, Instrumentation, and Control Engineering*, *2*(1), 1–4.

Demmers, T. G., Cao, Y., Parsons, D. J., Gauss, S., & Wathes, C. M. (2012). Simultaneous Monitoring and Control of Pig Growth and Ammonia Emissions. *IX International Livestock Environment Symposium (ILES IX)*. Valencia, Spain: American Society of Agricultural and Biological Engineers. DOI: 10.13031/2013.41560

El Bilali, A., Taleb, A., & Brouziyne, Y. (2021). Ground water quality forecasting using machine learning algorithms for irrigation purposes. *Agricultural Water Management, 245*, 106625. DOI: 10.1016/j.agwat.2020.106625

Galvão, R. K. H., Araújo, M. C. U., Fragoso, W. D., Silva, E. C., José, G. E., Soares, S. F. C., & Paiva, H. M. (2008). A variable elimination method to improve the parsimony of MLR models using the successive projections algorithm. *Chemometrics and Intelligent Laboratory Systems, 92*(1), 83–91. DOI: 10.1016/j.chemolab.2007.12.004

Gu, Y. Y, James. W, McNicol. M, (1994). An Application of Belief Networks to Future Crop Production. *IEEE Conference on Artificial Intelligence for Applications, San Antonia, TX*. pp. 305.

Harvey, C. A., Rakotobe, Z. L., Rao, N. S., Dave, R., Razafimahatratra, H., Rabarijohn, R. H., Rajaofara, H., & MacKinnon, J. L. (2014). Extreme vulnerability of smallholder farmers to agricultural risks and climate change in Madagascar. *Philosophical Transactions of the Royal Society of London. Series B, Biological Sciences, 369*(1639), 369. DOI: 10.1098/rstb.2013.0089 PMID: 24535397

Helm, J. M., Swiergosz, A. M., Haeberle, H. S., Karnuta, J. M., Schaffer, J. L., Krebs, V. E., Spitzer, A. I., & Ramkumar, P. N. (2020). Machine Learning and Artificial Intelligence: Definitions, Applications, and Future Directions. *Current Reviews in Musculoskeletal Medicine, 13*(1), 69–76. DOI: 10.1007/s12178-020-09600-8 PMID: 31983042

Islam, N., Rashid, M. M., Wibowo, S., Xu, C.-Y., Morshed, A., Wasimi, S. A., Moore, S., & Rahman, S. M. (2021). Early Weed Detection Using Image Processing and Machine Learning Techniques in an Australian Chilli Farm. *Agriculture, 11*(5), 387. DOI: 10.3390/agriculture11050387

Isleib, J. (2021). Signs and Symptoms of Plant Disease: Is It Fungal, Viral or Bacterial? https://www.canr.msu.edu/news/signs_and_symptoms_of_plant_disease_is_it_fungal_viral_or_bacterial

Khaki, S., & Wang, L. (2019). Crop Yield Prediction Using Deep Neural Networks. *Frontiers in Plant Science*, *10*, 621. DOI: 10.3389/fpls.2019.00621 PMID: 31191564

Lampridi, M., Sørensen, C., & Bochtis, D. (2019). Agricultural Sustainability: A Review of Concepts and Methods. *Sustainability (Basel)*, *11*(18), 5120. DOI: 10.3390/su11185120

LeCun, Y., Bengio, Y., & Hinton, G. (2015). Deep learning. *Nature*, *521*(7553), 436–444. DOI: 10.1038/nature14539 PMID: 26017442

Li, N., Ren, Z., Li, D., & Zeng, L. (2020). Review: Automated techniques for monitoring the behavior and welfare of broilers and laying hens: Towards the goal of precision livestock farming. *Animal*, *14*(3), 617–625. DOI: 10.1017/S1751731119002155 PMID: 31566170

Liakos, K., Busato, P., Moshou, D., Pearson, S., & Bochtis, D. (2018). Machine Learning in Agriculture: A Review. *Sensors (Basel)*, *18*(8), 2674. DOI: 10.3390/s18082674 PMID: 30110960

Matis, J. H., Birket, T., & Boudreaux, D. (1989). An Application of the Markov Chain Approach to Forecasting Cotton Yields from Surveys. *Agricultural Systems*, *29*(4), 357–370. DOI: 10.1016/0308-521X(89)90097-8

Monisha Kaul, M., Robert, L., Hill, H., & Walthall, C. (2005). Artificial neural networks for corn and Soybean yield prediction, *Elsevier. Agricultural Systems*, *85*(1), 1–18. DOI: 10.1016/j.agsy.2004.07.009

Nassani, A. A., Awan, U., Zaman, K., Hyder, S., Aldakhil, A. M., & Abro, M. M. Q. (2019). Management of natural resources and material pricing: Global evidence. *Resources Policy*, *64*, 101500. DOI: 10.1016/j.resourpol.2019.101500

Neupane, J., & Guo, W. (2019). Agronomic Basis and Strategies for Precision Water Management: A Review. *Agronomy (Basel)*, *9*(2), 87. DOI: 10.3390/agronomy9020087

Pan, S. J., & Yang, Q. (2010). A survey on transfer learning. *IEEE Transactions on Knowledge and Data Engineering*, *22*(10), 1345–1359. DOI: 10.1109/TKDE.2009.191

Papageorgiou, E. I., Aggelopoulou, K., Gemtos, T. A., & Nanos, G. D. (2018). Development and Evaluation of a Fuzzy Inference System and a Neuro-Fuzzy Inference System for Grading Apple Quality. *Applied Artificial Intelligence*, *32*(3), 253–280. DOI: 10.1080/08839514.2018.1448072

Samuel, A. L. (1959). Some Studies in Machine Learning Using the Game of Checkers. *IBM Journal of Research and Development*, *44*(1.2), 206–226. DOI: 10.1147/rd.441.0206

Seeland, M., Rzanny, M., Alaqraa, N., Wäldchen, J., & Mäder, P. (2017). Plant species classification using flower images—A comparative study of local feature representations. *PLoS One*, *12*(2), e0170629. DOI: 10.1371/journal.pone.0170629 PMID: 28234999

Sehgal, G., Gupta, B., Paneri, K., Singh, K., Sharma, G., & Shroff, G. (2017). Crop Planning using Stochastic Visual Optimization. *arXiv preprint arXiv:1710.09077*. DOI: 10.1109/VDS.2017.8573443

Senthil, K. S. D., & Shamili, M. D. (2022).Smart farming using Machine Learning and Deep Learning techniques, *Decision Analytics Journal,* Vol. 3,ISSN 2772-6622,DOI: 10.1016/j.dajour.2022.100041

Shaikh, T. N., & Mukane, S. M. (2017). Segmentation of Pomegranate Leaf for Detection of Disease Using Image Processing. *International Journal of Innovative Research in Computer and Communication Engineering*, *5*(4), 378–384.

Slaughter, D. C., Giles, D. K., & Downey, D. (2008). Autonomous robotic weed control systems: A review. *Computers and Electronics in Agriculture*, *61*(1), 63–78. DOI: 10.1016/j.compag.2007.05.008

Smola, A. (1999). Regression Estimation with Support Vector Learning Machines. *Master's Thesis, The Technical University of Munich, Munich, Germany,* pp. 1–78.

Sonka, S. (2016). Big Data: Fueling the Next Evolution of Agricultural Innovation. *J. Innov. Manag.*, *4*(1), 114–136. DOI: 10.24840/2183-0606_004.001_0008

Sørensen, C. A. G., Kateris, D., & Bochtis, D. (2019). ICT Innovations and Smart Farming. In *Communications in Computer and information science* (Vol. 953, pp. 1–19). Springer.

Stathakis, D., Savin, I., & Negre, T. (1994). Neuro-fuzzy modeling for crop yield prediction. *The International Archives of the Photogrammetry, Remote Sensing and Spatial Information Sciences*, *34*, 1–4.

Su, W. H. (2020). Advanced Machine Learning in Point Spectroscopy, RGB- and Hyperspectral-Imaging for Automatic Discriminations of Crops and Weeds: A Review. *Smart Cities*, *3*(3), 767–792. DOI: 10.3390/smartcities3030039

Suykens, J. A. K., Van Gestel, T., De Brabanter, J., De Moor, B., & Vandewalle, J. (2002). *Least Squares Support VectorMachines*. World Scientific. DOI: 10.1142/5089

Thayer, A., Vargas, A., Castellanos, A., Lafon, C., McCarl, B., Roelke, D., Winemiller, K., & Lacher, T. (2020). Integrating Agriculture and ecosystems to find Suitable Adaptations to Climate Change. *Climate (Basel)*, *8*(10), 10. DOI: 10.3390/cli8010010

Uno, Y., Prasher, S. O., Lacroix, R., Goel, P. K., Karimi, Y., Viau, A., & Patel, R. M. (2005). Artificial Neural Networks to Predict Corn Yield from Compact Airborne Spectrographic Imager Data. *Computers and Electronics in Agriculture*, *47*(2), 149–161. DOI: 10.1016/j.compag.2004.11.014

Van Klompenburg, T., Kassahun, A., & Catal, C. (2020). Crop yield prediction using machine learning: A systematic literature review. *Computers and Electronics in Agriculture*, *177*, 105709. DOI: 10.1016/j.compag.2020.105709

Veenadhari, S., Mishra, B., & Singh, C. D. (2011). Soybean Productivity Modelling using Decision Tree Algorithms. *International Journal of Computer Applications*, *27*(7), 975–8887. DOI: 10.5120/3314-4549

Yvoz, S., Petit, S., Biju-Duval, L., & Cordeau, S. (2020). A framework to type crop management strategies within a production situation to improve the comprehension of weed communities. *European Journal of Agronomy*, *115*, 126009. DOI: 10.1016/j.eja.2020.126009

Zecca, F. (2019). The Use of the Internet of Things for the Sustainability of the Agricultural Sector: The Case of Climate Smart Agriculture. *Int. J. Civ. Eng. Technol.*, *10*, 494–501.

Zhang, S., Huang, W., Huang, Y., & Zhang, C. (2020). Plant species recognition methods using leaf image: Overview. *Neurocomputing*, *408*, 246–272. DOI: 10.1016/j.neucom.2019.09.113

Zheng, A., & Casari, A. (2018). *Feature Engineering for Machine Learning: Principles and Techniques for Data Scientists*. O'Reilly Media, Inc.

Chapter 4

Challenges and Strategies for Management of Problematic Soils and Waters

Sachin Kumar
https://orcid.org/0009-0008-4604-8700
Chaudhary Charan Singh Haryana Agricultural University, Hisar, India

Pankaj Dahiya
https://orcid.org/0000-0002-7625-5451
Lovely Professional University, Phagwara, India

Satender Kumar
CCS Haryana Agricultural University, Hisar, India

Nisha
Kurukshetra University, Kurukshetra, India

Ravi Verma
https://orcid.org/0000-0001-8920-4867

DOI: 10.4018/979-8-3693-4392-0.ch004

Lovely Professional University, Phagwara, India

ABSTRACT

Problematic soils are a global concern due to the rising population and shrinking cultivable land, decreasing by 1-2% per year. This threatens the productivity of land needed for food security. Around 1125 million hectares are salt-affected, with 76 million hectares impacted by human activities like salinization, sodification, and acidification. The causes include high temperature, low rainfall, poor irrigation, and harmful agricultural practices like excessive tillage and fertilizer use. These degrade soil, making it non-productive. Advances in soil remediation, including biochar, water-saving irrigation, precision fertilization, and afforestation, offer solutions. Sustainable management strategies like reduced tillage and raised beds improve soil structure, water, and nutrient efficiency. Chemical and biological reclamation, including the use of salt-tolerant plants, can transform problematic soils into productive assets, enhancing food security and supporting farmers' incomes. Sustainable management is crucial for restoring soil health and agricultural productivity.

1.1 INTRODUCTION

A global phenomenon is the buildup of excess salts in the root zone, which can lead to a partial or total loss of soil productivity. All the continents have varying levels of soil salinity and sodicity, which pose a serious threat to agriculture. These soils make up about 7% of the world total land area and are most prevalent in arid and semiarid regions of the planet. Several factors including climate, landscape type, agricultural activities, irrigation techniques, policies pertaining to land management and so on, influence the distribution of salt-affected soil around the world.Saline soil is thought to cover 424 million hectaresof top soil (0-30 cm) and 833 million hectares of subsoil (30-100 cm), according to estimates from the Food and Agriculture Organization (FAO, 2021a) of the United Nations. Saline soil makes up 2.956 mha of the 6.73 mha of total salt-affected soils; the remaining 3.771 mha are sodic. The salt-affected soils are divided into the following states: Gujarat (2.23 mha), Uttar Pradesh (1.37 mha), Maharashtra (0.61 mha), West Bengal (0.44 mha) and Rajasthan (0.38 mha) (Kumar & Sharma, 2020).

Problematic soils are found in large areas as isolated patches scattered throughout regular soils; they do not form a continuous pattern and their quality decreases. Because of their detrimental effects on ecosystems, the environment, agricultural productivity and sustainability, soil salinity and sodicity are among the world's most significant salt-affected soils (SAS). Thus, it has a negative impact on soil and water quality, crop productivity, plant growth and ultimately, soil erosion and land degradation. The national economy is also impacted by salt-affected soils, in addition to the environment. The term "soil salinity" describes the existence of salts that dissolve in water, such as sodium (Na^+), potassium (K^+), chloride (Cl^-) and sulphate (SO_4^{2-}). While Na^+ and Cl^- are not regarded as plant nutrients, some ions, like K^+ and SO_4^{2-}, do function as plant nutrients. As a result, Na^+ and Cl^- are frequently the focus of soil salinity. An excess of Na^+ among the exchangeable cations in the soil solution is referred to as sodicity (Qadir et al., 2007). Salinity and sodicity have a direct impact on plant growth by affecting plant uptake of water and nutrients, as well as by imposing plant toxicity and, indirectly, by deteriorating soil physical conditions

1.2 SALINE SOIL

The amount of crop yield reduction is dependent on several variables, including crop growth, soil salinity, climate, etc. Crop growth may be completely stopped in extreme situations where there is an extremely high concentration of salts in the root zone. The excess salts in the root zone need to be eliminated to enhance crop growth in such soils. The process of eliminating soluble salts from the root zone is referred to as "reclamation" of saline soils. Frequently used or suggested strategies to achieve this include the following:

1.2.1 Scrapping

The process of scrapping involves removing salt-accumulated on topsoil over small land area to reduce salinity in root zone and its temporary toxic effects (Yuvaraj et al., 2021). Despite being a common practice among farmers, mechanically eliminating the salt buildup on the soil surface has not yielded significant results. Salt disposal is still a significant issue, even though this method may momentarily increase crop growth.

Figure 1. Diagrammatic representation of the formation of salt seeps (Stavi et al., 2021)

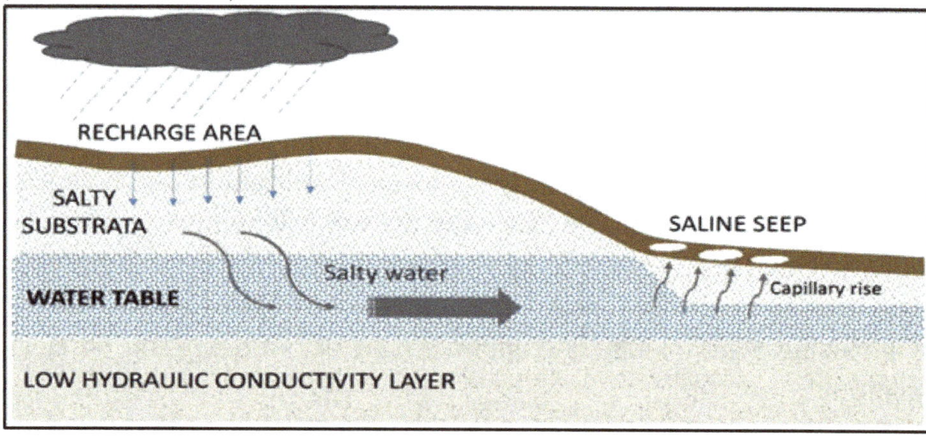

1.2.2 Flushing

Flushing is a useful technique for regulating the amount of salt present in the growing medium, which has an impact on ion uptake, water flow through the root and ultimately on plant quality (Bañón et al., 2019). This method has limited practical significance because only small amount of salts can be flushed from soil surface. A considerable amount of salts was not removed by sheet flow when water was passed over the surface. Less than 0.1% of the salts in the top 0.3 m of a silty clay loam soil were eliminated by flushing method (Reeve et al., 1955).

1.2.3 Leaching

In terms of eliminating salts from soil roots, this is the most efficient method available. It is achieved by accumulating fresh water on the soil's surface and letting it seep in. When subsurface drains are used to remove the leached salts from the reclamation area, it is an effective method of salt removal. In situations when there is adequate natural drainage but no artificial drains, leaching may lower salinity levels, *i.e.*there are no water table rises as the ponded water drains. Preferably, leaching should take place when the groundwater table is deep and the soil moisture content is low. Summertime evaporation losses cause significant amounts of water to be lost, making leaching generally less successful. However, the decision will rely on several factors, including the availability of water. Leaching, works well in certain regions of India during summer when the soil is dry and the water table is at its lowest point. Furthermore, significant volumes of fresh water can only be diverted during this period for reclamation purpose. The leaching fraction (LF) for a specific location can be computed using the irrigation and drainage water depth values (US Salinity Laboratory Staff,1954):

$$LF = D_{dw} / D_{iw} = Cl_{iw} / Cl_{dw} = EC_{iw} / EC_{dw}$$

Where, the drainage and irrigation water depths are denoted by D_{dw} and D_{iw}, respectively, concentration of Cl^- in drainage (Cl_{dw}) and irrigation (Cl_{iw}) water (Hoffman et al., 1979), ratio of EC_{iw} / EC_{dw} denotes the salinity in irrigation and drainage water (Rhoades and loveday, 1990).

1.2.3.1 Amount of water required for leaching

An accurate estimation of the amount of water needed to achieve salt leaching is crucial. Much of the water required for reclamation depends on several factors, including the original salt content of the soil, the desired level of salinity after leaching, the desired depth of reclamation and soil characteristics. Approximately 80% of salts in a unit depth of soil can be eliminated by a unit depth of water, which is a helpful generalization. As a result, 80% of the salts in the top 30 cm of soil will be eliminated by 30 cm of water passing through the soil.

Net water requirement = ET / 1-LR

Where, ET is the crop evapotranspiration demand and LR is expressed as a fraction (Ayers and Westcot 1985)

1.2.4 Water application technique

Salinity can be controlled by adjusting irrigation techniques, frequency and water application volume. Basin flooding, furrow irrigation, sprinkling and sub-surface irrigation are the primary methods of applying water. Unsaturated conditions were achieved in the field during leaching by using intermittent ponding or intermittent sprinkling at ratelower than the soil infiltration rate. (Nielsenet al., 1966), demonstrated that 75 cm of ponded water and 25 cm of sprinkled water had the same effect on lowering the salinity of the top 60 cm of soil.

1.2.4.1 Watering interval

To sustain a favorable salt balance in the root zone, irrigation in saline environments should satisfy both the plant water needs and leaching requirement. The greater the transpiration rate and the wetter the soil, the greater will be differences. This suggests that reducing irrigation intervals to keep the soil wet could potentially amplify the effects of salinity. Conversely, longer irrigation intervals typically lead to deeper roots and a higher percentage of water extraction from deeper zones. In saline conditions, there is a decrease in water uptake and consequently in ET losses. Soil with higher salinity will hold onto more water between irrigation than soil with lower salinity (Shainberg and Shalhevet 2012).

1.2.5 Crop management

The ability of plants to thrive and produce a satisfactory crop yield in saline soils varies greatly. Planning cropping schedules for maximum returns requires knowledge of the relative tolerance of different crops to a saline soil environment. Farmers may have to deal with salinity issues in certain circumstances, one such scenario is when their only irrigation water source is saline. The center of the raised bed or ridge has the highest concentration of salt accumulation, planting the seed there may have an impact on germination. Sloping beds can be used to better control salinity by planting seeds on the sloping side slightly above the water line. The benefits of alternate furrow irrigation include the ability to move salts beyond the single-seed row. The seeds are planted in a wet, less salinized zone at the bottom of the furrows, which promotes germination. Application of straw and polythene mulch reduces soil surface evaporation, which lowers the concentration of salt in the root zone profile.

Figure 2. Schematic representation of seed germination in furrow irrigation crops and salt movement (Bernstein & Fireman, 1957)

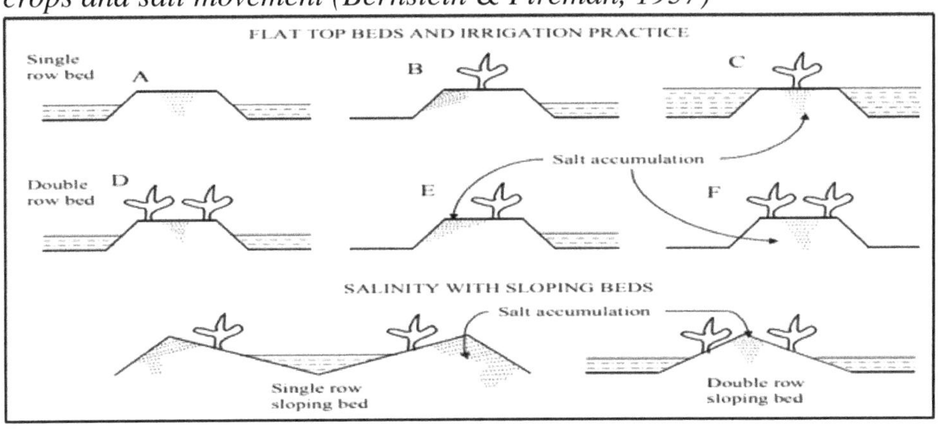

1.2.5.1 Varietal differences in salt tolerance

While it has long been known that different crop varieties have varying levels of tolerance to salinity and other unfavorable soil conditions, significant efforts to use these genetic differences in different breeding programs to maximize the potential of salt-tolerant crop varieties have only recently been made. In India and other coastal regions where sea inundations and seawater intrusion through rivers, estuaries and other channels pose a constant salinity challenge, rice has long been grown.

Numerous genotypes that are exceptionally salt tolerant were found after screening a wide variety of rice germplasm collected at various saline areas in India (Bhattacharya, 1976).

Table 1. Salt tolerance of crop based on threshold ECe (Grieve et al., 2012).

Common name	Tolerance based on	Threshold (ECe) (dS/m)
Rye	Grain yield	11.4
Canola	Seed yield	9.7
Guar	Seed yield	8.8
Barley	Grain yield	8.0
Cotton	Seed yield	7.7
Sugarbeet	Storage root	7.0
Sorghum	Grain yield	6.8
Triticale	Grain yield	6.1
Wheat	Grain yield	6.0
Cowpea	Seed yield	4.9
Sunflower	Seed yield	4.8

The salt tolerances of different crops are conventionally expressed as relative yield (Y_r), threshold value of salinity (a) and percentage decrease value per unit increase in salinity above the threshold value (b). Soil salinity is expressed in terms of ECe, in dS/m (Maas and Hoffman, 1977)

$$Y_r = 100 - b\,(ECe - a)$$

1.2.5.2 Growth stage

Poor and spotty stands and bare spots incultivated fields are typically caused by germination failures, though some crops appear to tolerate salinity well during seed germination as well as during later growth stages. This is often due to an abnormally high concentration of salt in the shallow surface zone where seeds are planted, rather than crops being particularly sensitive during germination. The salt left behind after the evaporation losses from soil surface is the cause of these high concentrations of salt. During germination, more plants are sensitive to salinity than at any other stage of growth. The sensitivity of germinating seeds to salinity varies greatly, though.

1.2.6 Mulching

Salts that have been leached tend to return to the surface during the times of high evapo-transpiration in between irrigations and during fallow conditions. Particularly high soil salinization occurs when there is a shallow water table and highly salinized groundwater. Controlling root zone salinity can be achieved through any techniques that lessen soil surface evaporationor promote soil water flow downward.To reduce salinity, bare fallow fields needed to have a soil mulch maintained (Sandoval and Benz (1966). Comparedto flooding or sprinkling bare soil, periodic sprinkling of mulched soils produced higher leaching efficiency due to greater removal of salt (Carter and Fanning, 1964).

1.2.7 Land levelling

To control salts, land leveling is crucial. Lack of leveling or uneven leveling wastewater and causes salts to accumulate beneath high spots and ridges.Certain places that are referred to as "hard spots" or "dry spots" form where crop growth is restricted. Reduced chances of maximum production can result from low spots that retain water for extended periods of time. Minor variations in the surface roughness of leveled land can lead to asymmetrical dispersal of water and salt, ultimately necessitating a multi-management initiative.

1.3 RECLAMATION OF SODIC SOIL

1.3.1 Use of Amendments

Most of the exchangeable sodium in the root zone must be removed to improve sodic soils and its place must be taken by calcium ions. There are several ways to do this; the most effective ones depend on the resources that are available, the local environment and the types of crops that will be planted on the reclaimed soils. A rice crop and other sodic-tolerant crops should ideally be included in the cropping sequence, along with the addition of organic residues and farmyard manure. Soil can still be reclaimed, but at a slower rate, by long-term irrigated cropping, provided the cultivator has very little money to spend on reclamation, the amendments are costly or unavailable and he is willing to wait many years before he can get good crop yields. Application of chemical soil amendments must apply before cropping to remove salts from the sodic soil through leaching. Soil amendments such as, sulphuric acid and sulphur that indirectly make the relatively insoluble calcium carbonate commonly found in sodic soils available for replacement of sodium through chemical or biological action. Materials like gypsum or calcium chloride directly supply soluble calcium for the replacement of exchangeable sodium. The dissolution of calcium compounds present in most soils is aided by the decomposition of organic matter and the root action of plants, although this process is comparatively sluggish.

1.3.1.1Kind of amendments (Abdel-Fattah 2019)

a. Soluble calcium salts: calcium chloride, gypsum.
b. Calcium salts of low solubility: ground limestone.
c. Acids or acid forming substances: sulphur, sulphuric acid, lime-sulphur, iron sulphate, pyrite, aluminium sulphate etc.

Several soil and financial factors will play a major role in determining which amendment is best for reclamation of sodic soil. Because limestone solubility rapidly decreases as soil pH increases, limestone ($CaCO_3$) is only a useful amendment in soils with pH levels below or around 7.0. It is evident that at pH values higher than 7.0, limestone efficacy as an amendment is significantly reduced.

1.3.1.2 Choice of amendment

The selection of an amendment in any given location will be determined by its relative efficacy, as determined by improvements in soil properties, crop growth and associated costs. When selecting an amendment, one must also consider how long it will take for the amendment to react in the soil and successfully replace adsorbed sodium. The most readily available source of soluble calcium is calcium chloride, which is highly soluble in water. However, due to its high cost, it has not been extensively used for reclamation. Like this, iron and aluminum sulphates are typically too expensive and have never been applied extensively to improve sodic soils. Relatively slow acting amendments include sulphur and pyrite, which must be converted to sulphuric acid by soil microorganisms before they can be used in reactions. Gypsum is the most used amendment because it is the least expensive and most readily available. In certain regions of the world, especially the western United States and some areas of the Soviet Union, sulphuric acid has also been widely used. The total amount of sodium that needs to be replenished determines how much of an amendment is required to recover sodic soil. This in turn relies on variables like the intended crops, the degree of soil deterioration as determined by the exchangeable sodium percentage (ESP), the soil texture and the mineralogical composition of the clay. The soil depth at which excess adsorbed sodium needs to be replaced for adequate crop growth is primarily determined by a crop's relative tolerance to exchangeable sodium and its typical rooting depth.

1.3.1.3 Application method

Broadly applied amendments, such as gypsum, are subsequently ploughed or disked into the soil. El-Gabaly (1971) found that gypsum applied on the soil surface was less effective than gypsum mixed with the top 15 cm of soil in terms of removing exchangeable sodium. Small amounts of gypsum at shallower depths proved to be more advantageous than combining at deeper depths. Lower depths caused the gypsum to become diluted, which lessened the depth-dependent decline in exchangeable sodium percentage (ESP). Additionally, a higher probability of a portion of the gypsum being utilized to neutralize soluble carbonates throughout the entire 30 cm soil depth at the expense of exchangeable sodium replacement at the shallower soil depth exists when gypsum is mixed to greater soil depths. This will lower the yield by decreasing the rate of seed germination.

1.3.2 Gypsum requirement

The gypsum requirement (GR) is the amount of gypsum needed to recover a sodic soil. The following formula can also be used to determine the gypsum requirement:

GR = (Present ESP - Desired ESP) x CEC x 0.021

When CEC is expressed in units of meq/100 g, the factor of 0.021 is used. Factor 0.0021 applies if CEC is expressed in mmols($^{+}$charge)/kg. These calculations assume that the irrigation water has a desired SAR and 90% reclamation efficiency (Robbins and Gavlak 1989).

1.3.2.1 Gypsum fineness and solubility

Gypsum is extracted from mine sites as lumps that must be ground before being applied to recover sodic soil. Economic considerations determine the required level of fineness for gypsum grinding. Though more expensive, very fine grinding is necessary due to physico-chemical factors (Aylmore et al., 1971), it is frequently believed that gypsum particles with a higher particle size will likely have a greater potential for sodic soil reclamation. The application of very finely ground gypsum to soil produced a high initial hydraulic conductivity that rapidly dropped over time (Chawla and Abrol (1982) with a highly sodic soil containing free sodium carbonate. Due to exchange sites favoring divalent calcium ions over sodium ions, gypsum becomes much more soluble when combined with highly sodic soil.

1.3.2.2 Gypsum-bed technology

To reduce the risk of sodicity in areas that are irrigated with RSC water, gypsum must be used. Gypsum can be mixed into the soil or placed in an irrigation channel (in gunny bags) so that it will gradually dissolve when water from the tube well falls. Better outcomes, though, are obtained when gypsum dissolving beds designed for this purpose are used (Tyagi, 2003). The gypsum-bed method involves running irrigation water through a chamber made of brick and cement that holds gypsum clods.

The rate of tube well discharge and the RSC of irrigation water determine the chamber's size. This chamber has connections to both a water channel and a water fall box on opposite sides. Ten centimeters up from the chamber's bottom, a net of iron bars covered in wire mesh (2 mm x 2 mm) is installed. Farmers can also turn their tube well chamber into a gypsum chamber by making the appropriate modifications. The gypsum placed in the chamber is dissolved and reclaimed by the sodic water that is flowing from below (Sharma, 2008).

1.3.3 Irrigation

The infiltration properties of the soil should determine the irrigation depth in sodic soils, rather than the traditional standard of the soil storage capacity. Several variables, including the amount of water evaporative demand in the atmosphere, the depth and proliferation of root penetration, soil ability to store and transfer water, irrigation interval. Due to poor root penetration, a reduced ability to hold water in an accessible form and poor transmission characteristics, sodic soils necessitate more frequent irrigations than those needed for crops grown in normal soil conditions. The degree of soil degradation will be the primary determinant of the ideal frequency. Sprinkler irrigation was preferable to surface water application because it was simpler to use and allowed for more frequent, light irrigations (Yadav and Girdhar 1977).

1.3.4 Mulching

Mulching can significantly aid in the leaching of soluble salts, lower exchangeable sodium percentage (ESP) and increase the yields of tolerant crops during the early years when the concentration of soluble salts is high in the surface soil layers.

1.3.5 Continuous cropping

Fallowing promotes the movement of salts upward. It is advised to continuously crop the land once sodic soil reclamation has begun. Sustaining cropping, especially with rice as one of the crops in the series, enhances the soil and eventually lowers exchangeable sodium percentage to a deeper depth. The submerged growing conditions of rice, which enable efficient leaching of exchange products formed during the growing season, are primarily responsible for the beneficial effects of rice cultivation.

1.3.5.1 Choice of crop

During the reclamation of sodic soils, crop selection is crucial. In the early stages of reclamation or when the crops are grown with irrigation water that poses a sodicity hazard, cultivating crops that are tolerant of excess exchangeable sodium can guarantee reasonable returns. Because excess exchangeable sodium negatively affects crop growth primarily through its adverse effect on soil physical properties, crops that can withstand conditions of excess moisture that result in short-term oxygen deficiencies are also more tolerant of sodic conditions. Compared to most field crops, grasses are generally more tolerant of sodic conditions. In soils with very high exchangeable sodium percentage (80 to 90), karnal grass grows exceptionally well even in the absence of amendments. Because of the biological activity of the grass roots, growing grasses causes the soil's physical qualities to improve over time and gradually decrease in sodicity. Soil improvement through the growth of tolerant grasses will not only increase soil absorption of precipitation and reduce runoff and erosion-related soil losses, but will also supply much-needed fodder.

Table 2. Sodicity tolerant crops based on exchangeable sodium percentage (Kaledhonkar et al., 2019)

ESP	Crops
10-15	Pea, safflower, banana, mash- bean
16-20	Papaya, citrus, soybean, Bengal gram
20-25	cowpea, guava, groundnut
25-30	Garlic, lemon grass, cotton, linseed
30-50	Wheat, ber, phalsa, mustard, sorghum
50-60	Barley, paragrass, sesbania, rhoades grass
60-71	Karnal grass, rice, sugarbeet

1.3.6 Organic manure

It has long been known that organic manures help sodic soils be reclaimed (Kanwar et al., (1965). Puttaswamygowda and Pratt 1973 attributed the positive effects of adding straw to a sodic soil under submerged conditions to: (i) the evolution of CO_2, the breakdown of organic matter and the release of some organic acids; (ii) the lowering of pH and the release of cations through the solubilization of $CaCO_3$ and other soil minerals, which increases the EC; and (iii) the substitution of Ca and Mg for exchangeable Na, which lowers the exchangeable sodium percentage. Organic materials have consistently shown benefits when applied in conjunction with inorganic amendments or in soils with mild sodicity; for this reason, their use is significant in the reclamation of sodic soils.

1.3.7 Laser Levelling

The most important prerequisites for managing sodic-water-irrigated land are leveling the ground and building bunds that are 30 to 40 cm high to collect and hold rainwater. It is also possible to shield the extremely erodible sodic soil on the surface from the pounding action of raindrops by ploughing the field in between downpours. In addition to augmenting rainfall intake, this approach aids in managing water losses due to evaporation and weed growth. Additionally, by allowing rain to dissolve the calcium carbonate in the soil, these methods encourage self-reclamation and uniform salt leaching.

1.4 RECLAMATION SALINE-SODIC SOILS

To restore the saline-sodic soils, it is crucial to first restore the sodic soil by using calcium to address the issue of elevated sodium concentration. By applying a large quantity of irrigation water, the issue of high salt concentration (salinity) can be remedied after the problem of high sodium concentration (sodicity) has been repossessed. Solubilizing sodium first, before leaching any other salts, is a fundamental requirement for saline-sodic soil reclamation. For this reason, even after treating the soil for salinity, problems of sodicity will persist if we do not make sodium soluble before eliminating all salts from the root zone. This will lead to a breakdown in the soilstructure, which will either completely halt or slow down the infiltration process. Remedial work becomes exceedingly challenging after this damage. Elevated electrical conductivity (EC) of irrigation water and soil facilitates soil structure enhancement, augments water penetration and prevents soil accumulation of sodium. High EC irrigation water has a beneficial effect on soil structure, but crop productivity suffers as a result. Like the treatment of sodic soils, the remediation of saline-sodic soils necessitates the addition of a step where calcium (Ca^{2+}) supplements, such as gypsum, are applied. This is followed by salinity remediation techniques that improve soil drainage and lower the water table.

1.5 RECLAMATION OF ACID SULPHATE SOILS

In the acid sulphate soils of the Sundarbans, it has been discovered that applying lime, rock phosphate and single superphosphate improves soil characteristics and increases crop yields (Burman et al., 2010). The following general tactics make up management measures for the potential and active acid sulphate soils.

1.5.1 Agronomic Measures

1.5.1.1 Selection of Tolerant Crops and Varieties

It has been reported that boosting organic acid biosynthesis by overexpressing genes derived from plants or bacteria that encode citrate synthase or malate dehydrogenase can improve the Al tolerance of *Arabidopsis thaliana*, Canola (*Brassica napus*), Alfalfa (*Medicago sativum*) and Tobacco (*Nicotiana tabacum*) (Panda et al., 2009). The rice varieties Canning 7, Mahsuri, SR 26B are advised for acid saline soils (pH 4.0–5.5). Fruit crops like citrus fruit, mango, watermelon and banana can be successfully grown under a raised-bed system (ridge and ditch system) in acid sulphate soils, as well as sugarcane and pineapple (Cho et al., 2002).

Table 3. Classification of crops based on soil acidity (Ray et al., 2015)

Crop	Acidophile (Soil pH: 4.5-5.5)	Tolerant crop (Soil pH: 5.5-7.0)	Sensitive crop (Soil pH: >7.0)
Cereal	Buck-Wheat, Oat, Paddy	Maize, Wheat, Barley	Millets
Pulses	Chickpea	Cowpea, Lentil, Rajma, Pigeon Pea	Pea
Oilseeds	Groundnut,	Sunflower, Sesame, Soyabean	Rapeseed And Mustard,
Vegetables	Potato, Radish, Brinjal, Sweet Potato	Beans, Broccoli, Cabbage, Cucumber	Bottle Gourd, Knolkhol, Ridge Gourd
Fruits	Citrus, Gooseberry, Pineapple	Peach, Watermelon, Papaya, Apple	Jackfruit, Banana
Flower	Rose	Merigold	Dahlia

1.5.1.2 Green/Organic Manuring and Mulching

Green manuring is the process of incorporating and ploughing in-situ green plants, especially legumes like *Crotalaria juncea* (sunnhemp), Dhaincha (*Sesbania* sp.) etc. to enhance the physical conditions, productivity and fertility of the soil (Rahman et al., 2021). This adds nitrogen and organic matter (OM) to the soil, increasing its fertility. The structure, microbial activity and water-holding capacity of the soil are all enhanced by the addition of organic matter (Morales 2022). Green manuring is done in the Ganges Delta prior to the wet/Kharif rice season. In this region, Dhaincha (*Sesbania culeate*) is the favored species for green manuring. It is sown in May or June using pre-monsoon rains and is ploughed during the puddling operations of Kharif rice in July, with a seed rate of 30-35 kg ha^{-1}. For optimal benefits, the soil should be amended with green manure crop while it is still in the vegetative stage. When the foliage is tender, which happens about 7-8 weeks after sowing, it should be turned into soil. Mature plants eventually become woody, so to allow the paddy to properly decompose, they must be buried before paddy transplantation (Sarangi et al., 2019). Applications of silica-rich materials, such as decomposed rice hull and poultry manure, can slow down the rate of P fixation during the crop-growing season (Thiyageshwari et al., 2018).

1.5.1.3 Application organic amendments

Application of FYM releases a wide variety of organic acids that form stable complexes with Fe and Al to block the P retention sites in acid soils where P fixation is a problem. This improves the availability and efficiency of P use (Agegnehu and Amede, 2017). In acidic soils, adding organic fertilizers has been successful in lowering Al phytotoxic levels, increasing yields. The primary mechanisms attributed to these enhancements are believed to be the creation of organo-Al complexes, which reduce the toxicity of Al, or the direct neutralization of Al due to the pH rise brought on by the organic matter.

1.5.1.4 Soil Profile Distribution during Land Modification

Reclamation of large areas of highly salinized coastal swamps and acid sulfate is hindered by the economic factor. Many hectares of mangrove land in Southeast Asia and Africa were further degraded and eventually abandoned because of the foolish reclamation techniques used to these potential coastal lands. These techniques included excising brackish water through dicing and the construction of fishponds. If adequate amounts of high-quality water and other necessary conditions for managing water quality are not met, the potential acid sulfate soils might not be disturbed; on the other hand, it might be preferable to use this land for other profitable endeavors like forestry, salt pans, biodiversity preservation, etc. Alternatively, other minimum disturbance reclamation techniques could be used, like flooded rice farming and the use of shallow fish ponds that could be appropriate for raising shrimp (prawns) in certain circumstances (Fanning, 1993).

1.5.1.5 Leaching

Leaching of the acid produced during oxidation is one of the ameliorative treatments for acid sulphate soils because applying limestone to neutralize all possible acidity would be highly costly. Instead, it may be more practical to promote pyrite oxidation and the ensuing acid leaching to counteract any residual acidity (dent 1986).

1.5.1.6 Maintaining a High-Water Table

Maintaining a high-water table in the soil can help control pyrite oxidation and the emergence of acute soil acidity and the results of this approach can be seen very quickly. Because the organic matter breaks down more quickly due to microbial activity, the soil pH rises and its acidity decreases because of this process (Inglett et al., 2005). In the paddy fields, where sufficient fresh water is used for irrigation, maintaining a high-water table is practically the best option (Surendran et al., 2021). The relatively less toxic and older acid soils that appear to be suitable for rice cultivation are greatly improved by improved agronomic and irrigation water management strategies (Tanaka and Navasero 1966). The crop and water productivity of these soils could be greatly increased by strict irrigation and drainage management based on science. Continuous submergence resulted in an increase in the concentrations of Fe^{2+} and Al^{3+} in the soil, but a decrease in the concentrations of Mn^{2+} and salinity.

1.5.2 Microbial Measures

Sulphate-reducing bacteria (SRB) can reverse pyrite oxidation by re-establishing reducing conditions within the bunded area. This is a natural process that can be economically advantageous if in situ microbial generation of the acid-neutralizing capacity is marked (Das and Das 2015). To initiate the process of reducing bacteria through the re-establishment of reducing conditions, submergence and the addition of biodegradable organic carbon (OC) are necessary. When SRB was added to compost at a rate of 10% of the weight, it was incubated at ±35°C for 4-5 days until bubbles formed on the compost's surface. After that, the soil was applied and the pH and sulfate levels of the soil were raised. A series of electron acceptors such as nitrate, Mn and Fe oxides and sulphate, accept the electrons produced by the microbial breakdown of the added organic matter in water-saturated, anaerobic conditions, leading to the reformation of pyrite and an increase in pH (Bucka et al., 2018). Microbes that chelate Al^{3+} ions and lessen Al toxicity, such as phosphate-solubilizing bacteria (PSB) and free-living, nitrogen-fixing bacteria, are inoculated to produce low-molecular-weight organic acids and phytohormones, such as indoleacetic acids (IAAs) (Panhwar et al., 2012). Due to the influence of phytohormones, a decrease in nutrient toxicity and other factors, the application of biofertilizers to rice crops grown in acid soil can improve plant growth and yield.

1.5.3 Chemical Measures

Different kinds of soil amendments may raise the pH of the soil and subsequently promote crop growth and development, depending on their effectiveness and availability (Ebimol et al., 2017). Chemical soil amendments such asslag, dolomite, rock phosphate, lime, basalt etc. are used to reclaim the acid soils(Shamshuddin et al., 2016). By increasing the pH of the soil, precipitating the toxic metals and increasing plant growth,liming materials such as magnesium carbonate, hydrated/slaked lime (calcium hydroxide), basic slag ($CaSiO_3$), dolomitic limestone (calcium and magnesium carbonate) and burnt/quick lime (calcium oxide) can counteract the toxic effects of Al, Fe and Mn (Rahman et al., 2018). Applying ferrous sulphate, which precipitates as FeS, to the soil can drastically lower the amount of H_2S in the soil solution (Bandyopadhyay et al., 2003). By using the Calcium Carbonate Equivalent (CCE), liming material neutralizing valueare compared. The CCE (Calcium Carbonate Equivalent) of dolomite, quick lime, hydrated lime, basic slag 109, 179, 136 and 86% respectively, are compared to the CCE of calcium carbonate 100% (Peters et al., 1996).

1.5.3.1 Liming

Liming is the process of adding materials containing carbonates, oxides, hydroxides and related compounds or complexes of Ca or Mg to soil to neutralize its acidity and raise its pH. We refer to these substances as agricultural lime. Because liming materials are comparatively insoluble, they are added well in advance of the following cropping season to facilitate their dissolution and release of enough Ca^{2+} or Mg^{2+} ions to balance the soil acidity. The exchangeable H^+ ions are replaced by Ca^{2+} or Mg^{2+} ions, raising the pH of the soil. The following factors determine whether liming will successfully neutralize soil acidity: (i) lime rate (ii) lime purity (iii) fineness of lime particle (iv) degree of mixing with soil

(Rodríguez et al., 2009). Calcite ($CaCO_3$) and dolomite ($MgCO_3$) are examples of carbonates and they are more manageable and less expensive than other liming materials. Consequently, the most common application for these materials is liming agricultural soils. Limestone and dolomite are example of secondary rocks or minerals that are extracted from ores to increase their reactivity in soil. The benefit of dolomite is that it contains magnesium in addition to calcium, despite its comparatively higher cost and slower rate of action. The addition of MnO_2 reduced the Fe^{2+} concentration; this effect was more noticeable in the pH 3–4 soil than in the neutral soil. Since liming materials have a very little and transient effect on potential acidity, applying them sparingly and at regular intervals may be more beneficial (Thampatti et al., 1998). The equation can be used to determine the lime requirement (LR) of soils for specific pH:

$$LR = CEC\ (V_2 - V_1) / 100$$

Where, V_1 is base saturation of soil and V_2 is expected value after liming(Quaggio et al., 1985).

1.5.3.2 Integrated Management

The application of lime, biofertilizers and organic manures in addition to chemical fertilizers is known as integrated nutrient management (INM) and is used in acidic soils (pH < 5.5) to improve soil health and increase crop yields (Sarkar et al., 2020). For high-yielding, proper nutrient management, dry-season rice crops are crucial part of crop management. Obtaining high yields, limiting fertilizer costs and maintaining or even improving soil fertility are the goals. In addition to using locally available organic plant nutrients, this can be accomplished by utilizing chemical fertilizers to their full potential. FYM, being an inexpensive and readily accessible source of organic nutrients, can play a significant role in Integrated Nutrient Management (INM) since cattle are a staple of the farming systems of most farmers in the Ganges Delta (Sarangi and Maji 2017). The free-floating water fern Azolla, another potential component of INM, fixes atmospheric nitrogen in a symbiotic relationship with the cyanobacterium Anabaena azollae (Rashad, 2021). The nitrogen fixed by the Anabaena species of blue-green algae (BGA) found in the lobes of Azolla leaves is present, its incorporation improves soil nutrients considerably. Common coastal ditches and ponds are ideal places for farmers to cultivate their own Azolla inoculums. For the most part, the climatic conditions for Azolla cultivation are favorable, as the mean temperature during the Boro season in most areas falls within the required range of 20-30°C.

1.5.4 Engineering Measures

By using neutralizing materials to buffer the acidic water that is already present and hydraulically suppressing the acid export from the soil (Powell and Martens 2005). Frequent tidal flooding of the low-lying regions keeps the soils moist, which inhibits pyrite oxidation and the production of more acid. Iron and aluminum compounds are precipitated from the site and buffering is increased when acid drainage water is treated with hydrated lime $Ca(OH)_2$. Sluicing or hydro-cycloning are the methods used to hydraulically separate the sulphides from the sand materials in coarse-textured soils containing iron sulphides (Osman 2018). This technique is appropriate for these types of soils. This method works very well in regions where the sedimenthas little organic matter and between 10 and 20% silt and clay. Special management is needed for the separated sulphide material extracted by the process, which could involve neutralization or strategic reburial (das and das 2015). Acid sulphate soil pH was raised, their acidity and aluminum concentrations were lowered by subsurface chemical treatments (injection through subsurface irrigation) using fine-grained $CaCO_3$ with a median grain size of 2.5 μm (Dalhem et al., 2019).

1.6 RECLAMATION OF SALINE WATER

The acceptable limits of salinity levels of water for irrigationhave wide variations. Therefore, it is important to create guidelines for each major area with similar soil, climatic and agricultural conditions, as any strict generalizations could be detrimental to workers at the field level. If water quality stays the same, management techniques can frequently be changed to achieve a more favorable distribution of salts in the profile and, consequently, higher crop yields. These management techniques can assist in resolving an irrigation water problem caused by excessive salinity. The process of desalinating water to eliminate soluble salts has frequently been mentioned as a potential technological advancement. However, given current technological advancements, it is unlikely that this approach will be widely used to utilize salty water for crop irrigation, at least not anytime soon.

1.6.1 Irrigation frequency

Frequent irrigation helps reduce the negative effects of the high salinity of irrigation water on the crops. Higher soil water contents in the upper portions of the root zone are maintained by more frequent irrigations, which also lower the concentration of soluble salts. These two elements work together to lessen the impact of elevated salt levels on plant water availability, which improves crop growth. Increased water application frequency is generally more compatible with the sprinkler irrigation method. However, increased irrigation frequency almost always results in a noticeable increase in water consumption when using surface irrigation techniques. To achieve optimal crop yield and water efficiency, it is important to carefully monitor the timing and quantity of water applied to the rootzone, particularly when using saline irrigation.

1.6.2 Use of extra water for leaching

It is essential to periodically remove salts from the soil by applying more water than is consumed to prevent excessive salt buildup. If the soil has sufficient internal drainage, the excess water applied will eliminate salts from the root zone. According to the crop's tolerance level, leaching requirement (LR) is the percentage of the total amount of water applied that needs to drain below the root zone to limit salinity to a particular level (Richards, 1954).

1.6.3 Selection of salt tolerant crops and varieties

While using high salinity water to grow a crop that is relatively sensitive to salt may be dubious, growing the right crop can yield good return even in this situation. Like this, there are plethora of opportunities to use saline water resources for crop production through the selection and breeding of crop varieties resistant to salt. Certain researchers have proposed that soaking seeds in salt solutions for a specific amount of time can induce salt tolerance, leading to higher yields in soils irrigated with saline water.

1.6.4 Conjunctive use of fresh and saline water

When crops require more evapotranspiration than they can receive, there are instances where high-quality irrigation water is available, but not in sufficient amounts. With these circumstances, combining high salinity water with high quality water to create medium salinity irrigation water for use all through the cropping season could be one strategy for achieving maximum crop production. As an alternative, irrigation with high-quality water could be employed during the more crucial growth stages, such as when the crop is comparatively more tolerant and the salinized water during germination. The optimal options must be determined by more research, considering factors such as soil salinity, critical growth stages in relation to crop tolerance, etc.

1.7 RECLAMATION OF SODIC WATER

1.7.1 Application of amendments

Irrigation water with a sodicity hazard could be improved by raising the soluble calcium status of the water, which would reduce the amount of sodium to the divalent cations and, consequently, its adsorption on the soil exchange complex. This is because the accumulation of sodium ions on the exchange complex is primarily to blame for poor soil physical properties. Soluble calcium salts applied to the water will also neutralize the bicarbonate and carbonate ions, lowering the water sodicity hazard. The extent of the modification that needs to be implemented, the manner and frequency of implementation, etc. are a few examples of the practical queries. Gypsum can be mixed into the soil or properly placed in the water channel to dissolve gradually in lumps. Although calcium chloride is fast acting and highly soluble in water, it is usually too costly. Acid-forming, also known as acidicamendments consist of pyrite, elemental sulphur and sulphuric acid. When sulfuric acid and soil carbonate combine, soluble calcium is released for sodium to be exchanged. To create sulphuric acid, soil bacteria must oxidize elemental sulfur and pyrite and then react with water. It may take several months for these materials to form appreciable amounts of suphuric acid. When acidic or acid-forming amendments are introduced to the soil, calcium carbonate needs to be present.

Table 4. Gypsum equivalent of amendments (Qadir et al., 2001)

Amendments	Chemical formula	Relative quantity
Ferrous sulphate	$FeSO_4.7H_2O$	1.61
Ammonium sulphate	$Al_2(SO_4)_3;18H_20$	1.29
Gypsum	$CaSO_4.\ 2H_2O$	1.00
Calcium chloride	$CaCl_2.2H_2O$	0.85
Calcium carbonate	$CaCO_3$	0.58
Sulphuric acid	H_2SO_4	0.57

These amounts are based on materials that are 100% pure. If there are any impurities in the material, the appropriate adjustments must be made. Consequently, if gypsum is only 80% pure, $1.0 \times (100 \div 80) = 1.25$, instead of adding 1.00 ton final quantity to be added is 1.25.

1.7.2 Irrigation interval

Water with a sodicity hazard can be effectively managed by frequent irrigation with small amounts of water. Water supply to the root is restricted by decreased soil permeability. Additionally, applying a lot at once may cause surface stagnation, which is bad for most crops. Regular irrigation could also lessen the amount of calcium that precipitates by reacting with bicarbonate in water, keeping the soil moist. Sprinkler irrigation with controlled water delivery interval should be taken into consideration.

1.7.3 Multi-Quality Water Resources Management

1.7.3.1 Mixing with an alternate source of water

Reducing the sodicity hazards of irrigation water on soil health and crop productivity can be achieved by combining the use of canal and sodic water. This is especially important in places where canal water supplies are either erratic or insufficient, as farmers frequently use sodic groundwater for additional irrigation. Sensitive crops can be grown in well-quality waters and tolerant crops can be grown in sodic water, for optimal utilization. Some conditions allow for the simultaneous or intermittent availability of both high- and low-quality waters.It is possible to blend different quality water within the supply network to provide tailor made water for every crop and soil type (Minhas and Gupta 1992).

1.7.3.2 Alternate use of poor and good quality water

Adjusting water sources throughout the growing season in accordance with the crucial stage of crop growth.Blending is a potentially useful technique in places where freshwater resources are available on demand. The amount of sodic and canal water mixed ensures that the crop to be grown has a final SAR/RSC below the threshold. The ratio of mixing canal and sodic water—two distinct water sourcesdepend on the crop ability to withstand sodicity.

1.7.4 Growing crops with low water requirement

Small amount of water should be used when irrigation water tends to cause sodicity issues. Water containing large amount of residual sodium carbonate (RSC) will continuously raise the exchangeable sodium status of soil, necessitating water use restrictions. Applying more water than necessary to meet evapotranspiration requirements is advised for saline water, but adding more water that poses a sodicity hazard will make the issue worse. When evapotranspiration demands are high, it is best to avoid planting crops and irrigating during those times.

1.7.5 Growing tolerant crops

Better yield can be achieved by cultivating crops that can withstand high exchangeable sodium level and unfavorable soil physical conditions, as opposed to sensitive crops. The physical attributes significantly improved because of the litter produced by *Prosopis juliflora*, *Terminalia arjuna*, *Casuarina equisetifolia*, *Pomacea pinnata* and*Acacia nilotica* after ten years of tree growth (Singh et al.,2008)

1.7.6 Organic matter application

Application of organic materials such as rice hulls, sawdust, sugar factory wastes, etc.; heavy dressing of organic manure; and regular incorporation of crop residues have been discovered to be helpful in preserving and enhancing the physical characteristic of soil and mitigating the negative effect of elevated exchangeable sodium levels. Therefore, if there is a risk of sodicity in the irrigation water, applying organic matter is especially advised wherever practical.

REFERENCES

Abdel-Fattah, M. K. (2019). Reclamation of saline-sodic soils for sustainable agriculture in Egypt. Sustainability of Agricultural Environment in Egypt: Part II: Soil-Water-Plant Nexus, 69-92.

Agegnehu, G., & Amede, T. (2017). Integrated soil fertility and plant nutrient management in tropical agro-ecosystems: A review. *Pedosphere, 27*(4), 662–680. DOI: 10.1016/S1002-0160(17)60382-5

Ayers, R. S., & Westcot, D. W. (1985). Water quality for agriculture (Vol. 29, p. 174). Rome: Food and agriculture organization of the United Nations.

Aylmore, L. A. G., Karim, M., & Quirk, J. P. (1971). Dissolution of gypsum, monocalcium phosphate and superphosphate fertilizers in relation to particle size and porous structure. *Soil Research (Collingwood, Vic.), 9*(1), 21–32.

Bandyopadhyay, B. K., Maji, B., Sen, H. S., & Tyagi, N. K. (2003). Coastal soils of West Bengal—Their nature, distribution and characteristics. *Bulletin, 1*, 1–62.

Bañón, S., Ochoa, J., Bañón, D., Ortuño, M. F., & Sánchez-Blanco, M. J. (2019). Controlling salt flushing using a salinity index obtained by soil dielectric sensors improves the physiological status and quality of potted hydrangea plant. *Scientia Horticulturae, 247*, 335–343. DOI: 10.1016/j.scienta.2018.12.026

Bernstein, L., & Fireman, M. (1957). Laboratoty Studies on Salt Distribution in Furrow-Irrigated Soil With Special Reference to the Pre-Emergence Period. *Soil Science, 83*(4), 249–264. DOI: 10.1097/00010694-195704000-00001

Bucka, F., Kölbl, A., Marschner, P., Fitzpatrick, R., Mosley, L., & Kögel-Knabner, I. (2018). Remediation of acid sulphate soils by addition of organic matter. In EGU General Assembly Conference Abstracts (p. 2378).

Burman, D., Bandyopadhyay, B. K., & Mahanta, K. K. (2010). Management of acid sulphate soil of coastal Sundarbans region: Observations under on-farm trial. *Journal of the Indian Society of Coastal Agricultural Research*, *28*(1), 8–11.

Carter, D. L., & Fanning, C. D. (1964). Combining surface mulches and periodic water applications for reclaiming saline soils. *Soil Science Society of America Journal*, *28*(4), 564–567. DOI: 10.2136/sssaj1964.03615995002800040033x

Chawla, K. L., & Abrol, I. P. (1982). Effect of gypsum fineness on the reclamation of sodic soils. *Agricultural Water Management*, *5*(1), 41–50. DOI: 10.1016/0378-3774(82)90037-3

Cho, K. M., Ranamukhaarachchi, S. L., & Zoebisch, M. A. (2002). Cropping systems on acid sulphate soils in the central plains of Thailand: constraints and remedies. 17th WCSS, 14-21.

Dalhem, K., Engblom, S., Stén, P., & Österholm, P. (2019). Subsurface hydrochemical precision treatment of a coastal acid sulfate soil. *Applied Geochemistry*, *100*, 352–362. DOI: 10.1016/j.apgeochem.2018.12.005

Das, S. K., & Das, S. K. (2015). Acid sulphate soil: management strategy for soil health and productivity. Popular kheti, 3(2), 2-7.

Dent, D. L. (1986). *Acid sulphate soils: a baseline for research and development (No. 39)*. ILRI.

Ebimol, N. L., Suresh, P. R., Binitha, N. K., & Santhi, G. R. (2017). Management of iron and aluminium toxicity in acid sulphate soils of Kuttanad. *International Journal of Current Microbiology and Applied Sciences*, *6*(11), 1496–1503. DOI: 10.20546/ijcmas.2017.611.177

El-Gabaly, M. M. (1971). Reclamation and management of salt affected soils. In *Salinity Seminar, Baghdad. Irrigation and Drainage Paper, 7*. FAO.

Fanning, D. S. (1993). Salinity problems in acid sulfate coastal soils. Towards the rational use of high salinity tolerant plants: Vol. 1 Deliberations about High Salinity Tolerant Plants and Ecosystems, 491-500. DOI: 10.1007/978-94-011-1858-3_52

Grieve, C. M., Grattan, S. R., & Maas, E. V. (2012). Plant salt tolerance. ASCE manual and reports on engineering practice, 71, 405-459.

Hoffman, G. J., Rawlins, S. L., Oster, J. D., Jobes, J. A., & Merrill, S. D. (1979). Leaching requirement for salinity control I. Wheat, sorghum and lettuce. *Agricultural Water Management*, *2*(3), 177–192. DOI: 10.1016/0378-3774(79)90001-5

Inglett, P. W., Reddy, K. R., & Corstanje, R. (2005). *Encyclopedia of soils in the environment.*

Kaledhonkar, M. J., Meena, B. L., & Sharma, P. C. (2019). Reclamation and nutrient management for salt-affected soils.

Kanwar, J. S., Bhumbla, D. R., & Sing, N. T. (1965). Studies on the reclamation of saline and sodic soils in the Punjab.

Maas, E. V., & Hoffman, G. J. (1977). Crop salt tolerance—Current assessment. *Journal of the Irrigation and Drainage Division*, *103*(2), 115–134. DOI: 10.1061/JRCEA4.0001137

Minhas, P. S., & Gupta, R. K. (1992). *Quality of irrigation water: assessment and management.* Indian Council of Agricultural Research.

Morales, L., Domínguez, M. T., & Fernández-Boy, E. (2022, May). Effect of the addition of organic amendments to C-poor agricultural soils on soil resistance against drought. In EGU General Assembly Conference Abstracts (pp. EGU22-5532).

Nielsen, D. R., Biggar, J. W., & Luthin, J. N. (1966). Desalinization of soils under controlled unsaturated flow conditions. In *Proceedings of the 6th International Congress on Irrigation and Drainage* (pp. 15-19).

Osman, K. T. (2018). *Management of Soil Problems.* Springer. DOI: 10.1007/978-3-319-75527-4

Panda, S. K., Baluška, F., & Matsumoto, H. (2009). Aluminum stress signaling in plants. *Plant Signaling & Behavior*, *4*(7), 592–597. DOI: 10.4161/psb.4.7.8903

Panhwar, Q. A., Radziah, O., Zaharah, A. R., Sariah, M., & Mohd Razi, I. (2012). Isolation and characterization of phosphorus solubilizing bacteria. *African Journal of Biotechnology, 11*, 2711–2719.

Peters, J. B., Schulte, E. E., & Kelling, K. A. (1996). Choosing between liming materials. University of Wisconsin--Extension.

Powell, B., & Martens, M. (2005). A review of acid sulfate soil impacts, actions and policies that impact on water quality in Great Barrier Reef catchments, including a case study on remediation at East Trinity. *Marine Pollution Bulletin, 51*(1-4), 149–164. DOI: 10.1016/j.marpolbul.2004.10.047

Puttaswamygowda, B. S., & Pratt, P. F. (1973). Effects of straw, calcium chloride and submergence on a sodic soil. *Soil Science Society of America Journal, 37*(2), 208–212. DOI: 10.2136/sssaj1973.03615995003700020016x

Qadir, M., Oster, J. D., Schubert, S., Noble, A. D., & Sahrawat, K. L. (2007). Phytoremediation of sodic and saline-sodic soils. *Advances in Agronomy, 96*, 197–247. DOI: 10.1016/S0065-2113(07)96006-X

Quaggio, J. A., Van Raij, B., & Malavolta, E. (1985). Alternative use of the SMP-buffer solution to determine lime requirement of soils. *Communications in Soil Science and Plant Analysis, 16*(3), 245–260. DOI: 10.1080/00103628509367600

Rahman, M. A., Lee, S. H., Ji, H. C., Kabir, A. H., Jones, C. S., & Lee, K. W. (2018). Importance of mineral nutrition for mitigating aluminum toxicity in plants on acidic soils: Current status and opportunities. *International Journal of Molecular Sciences, 19*(10), 3073. DOI: 10.3390/ijms19103073

Rahman, M. M., Uddin, S., Jahangir, M. M. R., Solaiman, Z. M., Alamri, S., Siddiqui, M. H., & Islam, M. R. (2021). Integrated nutrient management enhances productivity and nitrogen use efficiency of crops in acidic and charland soils. *Plants, 10*(11), 2547. DOI: 10.3390/plants10112547

Rashad, S. (2021). An overview on the aquatic fern Azolla spp. as a sustainable source of nutrients and bioactive compounds with resourceful applications. *Egyptian Journal of Aquatic Biology and Fisheries, 25*(1), 775–782. DOI: 10.21608/ejabf.2021.150205

Ray, S. K., Chatterjee, D., Saha, S., Kamble, K. H., Deka, B. C., Kendra, V., & Jharnapani, M. (2015). *A guide for soil nutrient management with special reference to Wokha. ICAR Research Complex for NEH Region.* Nagaland Centre.

Reeve, R. C., Pillsbury, A. F., & Wilcox, L. V. (1955). Reclamation of a saline and high boron soil in the Coachella Valley of California. Hilgardia;(United States), 24(4). DOI: 10.3733/hilg.v24n04p069

Rhoades, J. D., & Loveday, J. (1990). Salinity in irrigated agriculture. *Agronomy (Basel)*, (30), 1089–1142.

Robbins, C. W., & Gavlak, R. G. (1989). Salt-and sodium-affected soils.

Rodríguez, J. Á. C., Hanafi, M. M., Omar, S. R. S., & Rafii, Y. M. (2009). Chemical characteristics of representative high aluminium saturation soil as affected by addition of soil amendments in a closed incubation system. *The Malaysian Journal of Soil Science, 13*, 13–28.

Sandoval, F. M., & Benz, L. C. (1966). Effect of bare fallow, barley and grass on salinity of a soil over a saline water table. *Soil Science Society of America Journal, 30*(3), 392–396. DOI: 10.2136/sssaj1966.03615995003000030026x

Sarangi, S. K., & Maji, B. (2017). Sustainable rice cultivation in coastal saline soils: a case study Institute, India. In *Achieving sustainable cultivation of rice* (Vol. 2, pp. 101–136). Burleigh Dodds Science Publishing. DOI: 10.19103/AS.2016.0003.07

Sarangi, S. K., Maji, B., Digar, S., Burman, D., Mandal, U. K., Mahanta, K. K., & Mandal, S. (2019). Technologies for Cropping System Intensification in the Salt-Affected Coastal Zone of West Bengal, India Technical Bulletin ICAR-CSSRI, p. 28

Sarkar, A. K., Pattanayak, S. K., Singh, S., Mahapatra, P., Kumar, A., & Ghosh, G. K. (2020). Integrated nutrient management strategies for acidic soils. *Indian Journal of Fertilisers, 16*(5), 476–491.

Shainberg, I., & Shalhevet, J. (Eds.). (2012). *Soil salinity under irrigation: Processes and management* (Vol. 51). Springer Science & Business Media.

Shamshuddin, J., Panhwar, Q. A., Shazana, M. A. R. S., Elisa, A. A., Fauziah, C. I., & Naher, U. A. (2016). Improving the productivity of acid sulfate soils for rice cultivation using limestone, basalt, organic fertilizer and/or their combinations. *Sains Malaysiana*, *45*(3), 383–392.

Sharma, D. R. (2008). *Characterization of Sodic Water and Technologies for its Efficient Use*. Conjunctive Use of Canal and Groundwater.

Singh, Y. P., Sharma, D. K., Singh, G., Nayak, A. K., & Mishra, V. K. (2008). Alternate Land Use Management for Sodic Soils. Technical bulletin no. 2/2008.

Stavi, I., Thevs, N., & Priori, S. (2021). Soil salinity and sodicity in drylands: A review of causes, effects, monitoring and restoration measures. *Frontiers in Environmental Science*, *9*, 712831. DOI: 10.3389/fenvs.2021.712831

Surendran, U., Raja, P., Jayakumar, M., & Subramoniam, S. R. (2021). Use of efficient water saving techniques for production of rice in India under climate change scenario: A critical review. *Journal of Cleaner Production*, *309*, 127272. DOI: 10.1016/j.jclepro.2021.127272

Tanaka, A., & Navasero, S. A. (1966). Growth of the rice plant on acid sulfate soils. *Soil Science and Plant Nutrition*, *12*(3), 23–30. DOI: 10.1080/00380768.1966.10431192

Thampatti, K. C., Cherian, S., & Iyer, M. S. (1998). Influence of liming materials on soil acidity characteristics of an acid sulphate soil. *Journal of the Indian Society of Soil Science*, *46*(2), 296–299.

Thiyageshwari, S., Gayathri, P., Krishnamoorthy, R., Anandham, R., & Paul, D. (2018). Exploration of rice husk compost as an alternate organic manure to enhance the productivity of blackgram in typic haplustalf and typic rhodustalf. *International Journal of Environmental Research and Public Health*, *15*(2), 358. DOI: 10.3390/ijerph15020358

Tyagi, N. K. (2003). Managing saline and alkaline water for higher productivity. JW, Kijne, R., Barker, D., Molden (eds.). Water productivity in agriculture: limits and opportunities for improvement. CAB International, 69-87.

U.S. Salinity Laboratory Staff. (1954). Diagnosis and improvement of saline and alkali soils. In Richards, L. A. (Ed.), *USDA Agricultural Handbook 60*. USDA.

Yadav, J. S. P., & Girdhar, I. K. (1977). Salt translocation and crop growth under sprinkler and surface methods of irrigation in sodic soil [India]. *Indian Journal of Agricultural Sciences*, •••, 47.

Yuvaraj, M., Bose, K. S. C., Elavarasi, P., & Tawfik, E. (2021). Soil salinity and its management. Soil moisture importance, 1, 109.

Chapter 5
The Influence of Activated Water on the Yield and Safety on Potato Tubers in Western Siberia

Olga Anatolievna Pasko
https://orcid.org/0000-0001-8322-834X

National Open Institute St. Petersburg, Russia & Leningrad Branch of the Federal State Budgetary Institution Agrochemical Service of Russia, St. Petersburg, Russia

Nadezhda Anatolievna Lebedeva
https://orcid.org/0000-0003-4095-2631

International Personnel Academy, Germany

ABSTRACT

Three-year field tests of pre-planting treatment of potato tubers with electrochemically activated water were carried out. The maximum increase in the total tuber yield was obtained when using catholyte. Additional processing of vegetative potato plants increased the yield by 30%. The bactericidal effect of anolyte on potato tubers during winter storage was revealed. In the variant with the Lugovskoy variety, they were 8% less

DOI: 10.4018/979-8-3693-4392-0.ch005

affected by dry rot, and were not at all affected by wet rot (in the control - 14%). The overall incidence decreased by 20%. Activated water for pre-planting treatment of potato tubers and before planting them is an effective and environmentally friendly method of increasing yield and product quality.

1. INTRODUCTION

The range of applications of chemical stimulants is very diverse (Giri, Schillinger, 2003; Devika, Singh, et al., 2021). In particular, they allow to remove buds, tubers, bulbs, and seeds from a resting state (Anwar, Xianchang et al., 2020; Bradford, Steiner et al., 1990). Biologically active substances are used less frequently in potato growing than in cereals, vegetables fruit and berry crops, although the possibilities for regulating growth, development, and stimulation of productivity in this industry are much wider. With their help, the number of awakened eyes on the tuber can be increased, which affects the number of sprouts, stems, stem density, and, consequently, yield (Damalas, Koutroubas et al, 2019).

Stimulation of growth processes accelerates the germination of tubers, as well as the onset of the main phenological phases. More intensive development makes it possible to switch to autotrophic nutrition earlier, make better use of the reserves of spring productive lag, and soil nutrients and form a harvest at an earlier date. This makes the plant less susceptible to many diseases as a result of the formation of the bulk of the crop before the mass spread of diseases. Accelerated development contributes to better ripening of the peel of tubers, which ensures their better preservation during winter storage. It should be noted that growth regulators not only stimulate productivity but also increase the resistance of plants to 1environmental factors.

The biogenic group includes stimulants obtained from plant or animal tissues. As a rule, they have a complex and uncertain composition, but they are highly effective and have been used in practice for a long time. Effective growth regulators are sodium humates obtained from coals or peat (Pasko, 2011 c). Their use makes it possible to accelerate the onset of the main phases of development by 2-4 days, stimulates the development of the leaf surface, the synthesis of chlorophyll and, as a result, potato productivity. In addition, sodium humate (50-100 mg/l) contributes to the formation of greater plant resistance to frost. The stimulating properties of sodium humate are also shown for vegetable crops.

Their disease damage was either absent or exceptional. At the same time, in the control, all the fruits of the mat were severely affected by this disease (more than three points on a five-point rating scale). A study of the effect of pre-sowing (pre-planting) treatment of plant material with gamma irradiation showed that optimal stimulation occurred at doses of 100-500 rad. The increase in potato yield was 10-40%, and the increase in yield was mainly due to an increase in the mass of tubers. At the same time, the quality of tubers improved: the content of starch, vitamin C, protein, and amino acids increased. Similar results were published in Hungary and Italy. Interesting data were obtained by Zeynalova and Alieva (cited by Pasko & Gomboev, 2011). Potato blight disease decreased from 31-52% to 16-29% in long-term trials.

An important point turned out to be that the positive effect of irradiation persisted for at least three months of weathering in the storage. “Stimulated” tubers were better stored. According to the Leningrad Regional Plant Protection Station, treatment of tubers with copper preparations contributed to increased resistance of potatoes to diseases, better tuberization and provided an increase in yield by 2.4-2.7 t/ha (0.02% solution of copper sulfate of 20 litres per ton of tubers was used. The observations were carried out on an area of 14 thousand hectares). A valuable consequence of the pre-planting treatment of potato tubers is an increase in plant resistance to viral diseases. Two-hour treatment with solutions of indoleacetic acid (0.0005%), indole propionic acid (0.0005%) and proline (0.0001%) reduced the symptoms of viral diseases from 50.3% in the control to 20.9% when using indoleacetic acid, 27.1% indole propionic acid, 21.1% (more than twice) – proline (cited by Pasko, Semenov et al., 2007).

In the experiments of the Luga Agrochemical laboratory, dry beet seeds and seeds soaked for 24 hours in 0.1% boric acid solution were sown. The yield in the boric acid variant was twice as high (42.0 t/ha) as in the control. Pre-planting treatment of potato tubers with boric acid did not give a stable increase in yield: it increased only in one of the three years of testing. A solution of zinc sulfate (15-20 l/t of tubers) increased the resistance of potatoes to wet rot. Storage losses decreased in "Ogonek" varieties - from 4.2 to 1.6%, Gatchinsky - from 14.8 to 6.2%).

Pre-sowing treatment of feed bean seeds with molybdenum preparations gave an increase of 300 kg/ha (in the control - 2400 kg/ha). The use of 0.5% zinc sulfate solution for pre-planting stimulation of potato tubers of "Ogonek" and "Table" varieties increased yields by 11-12%. The increase in yield occurred not only in the year of processing but also in the following (by 18%). Pre-sowing treatment of wheat, barley, oats, and corn seeds with 2.0-2.5% sodium humate solution (200-250 g/t), provided additional production of 100-200 kg/ha, increased the dry residue content in grains (cited by Zalepukhin, Pasko et al., 1988).

Improving pre-sowing treatment is possible based on hot and cold spraying of biological products such as biological products potassium humate and sodium humate. According to Goryachkina, Zinovyev et al. (2022) in treated plants there is an increase in vegetative activity, accelerated plant development and an increase in the physiological activity of tubers. Similar results were obtained when using active iron nanopowders (Kovalenko & Folmanis, 2006), and plant growth regulators (Kefeli, 1978).

Because the action of factors of the most diverse nature is effective (i.e., to obtain the same result, you can always choose a certain dose of any factor), the criteria of environmental safety, cheapness and accessibility of the stimulant come to the fore (Pasko, Pat. 2375313 RU; Pasko, Semenov, Copyright certificate No. 161 19557; No. 16119557). In our opinion, the so-called activated water meets the complex of these requirements. In recent years, it has become widespread in agriculture. However, its practical application has always been ahead of theoretical research. Therefore, it is not uncommon for the existence of abnormal, biologically active properties of water to be questioned at all due to insufficient scientific evidence.

This causes increased interest among scientists of different specialities in studying the properties and practical application of the differences of its various modifications to accelerate technological processes in biology, agriculture, biotechnology and chemistry, as well as in medicine - for the prevention and treatment of certain diseases.

Production tests of pre-sowing treatment of *electrochemically activated water* (ECAW) agricultural seeds were carried out in 17 farms of

Kyrgyzstan, Uzbekistan, Ukraine, Kazakhstan and Russia (Volgograd, Leningrad, Kuibyshev, Tomsk regions, Stavropol, Krasnodar Territory and Tatarstan). The following crops were selected as the studied crops: corn, buckwheat, millet, sunflower, oats, wheat, barley, sweet pepper, tomato, cucumber, radish, lettuce, onion and cabbage. We consider such a growth stimulant as activated water to be the most environmentally friendly. Various assumptions are made about the causes of its abnormal properties (they will be discussed later), but chemically activated water is water with several altered physicochemical properties. After an incident of a certain time (relaxation period) activated water goes into an equilibrium (initial) state and does not pose any danger from an environmental point of view.

When the basic conditions of temperature and humidity parameters of the atmosphere in the storage are met, pre-planting treatment of tubers becomes a new resource. For example, the use of hydrogen peroxide in Heartland (New Brunswick, Canada) resulted in significantly higher percentages of seedling emergence and total and marketable yield compared to other treatments (Al-Mughrabi, 2007). Due to the low cost and efficiency of using hydrogen peroxide, this drug and technology are widely used in practice (Afek, Orenstein et al., 2001; Norikane, Brook et al., 2001; Olsen et al., 2000) along with traditional fungicides (Secor and Gudmestad, 1999).

We have previously shown that treatment of electrochemically activated water in amaranth seeds, selected as a model, from the first minutes led to a sharp increase in the number of free radicals in seed tissues and an increase in membrane permeability (Pasko & Ketskalo, 1991). This contributed to the acceleration of seed germination of all plant species we studied (about one hundred and twenty species) (Pasko, 1991), and an increase in germination energy (Pasko, 2011). Biochemical studies have revealed the initiation of alternative metabolic pathways due to temporary hypoxia, characteristic of catholyte and catholyte-anolyte mixtures (Pasko, 2009; 2011; Pasko & Gomboev, 2011a).

Similar conditions stimulated the processes of histogenesis in model objects for studying the effects of growth regulators on bean cuttings, their number of roots and their development increased (Pasko & Leonov, 2011). At further stages of plant development, the processes of their organogenesis and morphogenesis intensified (Pasko, 2000), which led to the enhanced development of the vegetative system first (Pasko, 2000a; 2010a; 2011a), and then the generative system (Pasko, 2010b). Regulating the water treatment regime and the physicochemical properties of activated water (Pasko, 2010c; 2024; Pasko, Semenov et al., 2007). From a scientific point of view, the control of the output parameters of activated water made it possible to achieve a given crop yield in large-scale experiments in production conditions (Pasko, 2011a; 2011b). Taking into account the characteristics of seeds and obtaining the necessary volumes of electrochemically activated water, various models of electrolyzers and methods of their application were developed and tested (Pasko, Semenov, 1991; 1993a; 1993b). During the study, with the expansion of the number of studied species, the attention of researchers was naturally attracted by the potato culture (Latin Solanum tuberosum).

To store tubers, storage facilities with artificial ventilation and optimal temperature and humidity conditions are used. This reduces losses due to shrinkage or wilting, germination of tubers, damage by pests, mould and other pathogens, and the development of rotting. It is important that once a problem is detected, it becomes almost impossible to stop the process.

Studies by Semenenko, Belitskaya et al. (2013) found an improvement in the phytosanitary condition of soils for soil-dwelling phytophages and beneficial mesofauna under the influence of pre-planting treatment of potato tubers with electrochemically activated water. Anolyte with ORP +500 mV and catholyte with ORP and -700 reduced the number of wireworms by 20-35% and inhibited the activity of this pest. The authors note an increase in the yield of tubers with an increase in the ORP of activated water, however, they do not provide a mathematical model of this dependence. The field research data are interesting – in the variant with an ORP of activated water equal to +300 mV, the yield increase reached 0.3 t/ha (1.8%) on average and 31.7% maximum (when using an anolyte with an ORP equal to +700 mV). Unfortunately, the authors do not provide statistical processing data, which does not allow us to judge the reliability of these changes.

2. OBJECTS AND METHODS OF RESEARCH

2.1. Geographical Location and Weather Conditions of the Tomsk Region

The Tomsk Region lies almost entirely in the southeastern part of the West Siberian Lowland, in a zone of risky agriculture, bordering the Kemerovo, Novosibirsk, Omsk, Tyumen regions and the Krasnoyarsk Territory. The elevations of the lowland do not exceed 200 m above sea level (Johannsen, 1971). The geographical boundaries of the region are as follows: from south to north - from 55.7 to 61.0° north latitude (600 km), from west and east - from 75.0° to 89.4 ° east longitude from Greenwich (850 km). The territory of the region is 314.4 thousand km^2 The taiga zone occupies huge areas - from the polar route to about 57 ° C., i.e. about 60% of the area of Western Siberia. It is heterogeneous. In the north, rare-coniferous spruce-deciduous swampy forests turn into deciduous cedar mixed with pine riverine forests, which occupy most of the zone. The Mesopotamians are swampy and practically treeless.

Further south is a zone of birch-pine-dark coniferous swampy forests. A narrow strip of small-leaved forests (birch, aspen with an admixture of pine) adjoins the forest-steppe zone. The soil cover of this subdistrict is represented by grey forest soils of all three subtypes, as well as sod-podzolic). Leached chernozems are less common. The sum of soil temperatures at a depth of 10 cm for a period with temperatures above 10 ° for clay and heavy loamy mechanical composition is 1600-1800 °, loamy - 1700-1900 °, sandy loam and sandy - 1800-2000 °. Swamp soils are gaining temperatures in the range of 1200 - 1400 °.

The soil composition of the experimental sites where the research was conducted is characterized by certain features. Thus, the experimental site on the territory of the Tomsk agricultural station is characterized by grey forest medium loamy soils with humus content in the range of 5.2-5.5%, nitrogen 0.25-0.30% and total phosphorus 0.14-0.17%. The amount of absorbed bases is 25-30 mg*eq/100 g of soil, and the pH of the salt extract is 5.9-6.0. The precursors of vegetable crops and potatoes are steam, perennial grasses, cereals, and flax. The soils of the Tomsk State Agricultural Testing Station are characterized by a slightly acidic reaction of the medium in the arable layer and have 6.3% humus, 0.3% total nitrogen, and 0.25% total phosphorus. The content of exchangeable phosphorus is 0.25%, potassium is 14 mg, and mobile phosphorus is 37.6 mg per 100 g of soil.

The experimental site of the Siberian Botanical Garden is located on grey forest medium loamy soils with humus content in the range of 3-4%, mobile forms of phosphorus and potassium 5-10 mg/100 g of soil. They are characterized by low water and air permeability, and a tendency to swim. The reaction of the soil solution is slightly acidic, it is 5.9-6.0.

2.2. Meteorological Conditions

According to the Agrometeorological Bulletin, the average annual air temperature in the experimental area is -0.6 ° C. The number of days in a year with temperatures above 5 ° is 153, above 10° is 113, and the sum of temperatures above 10° reaches 1750 °. The growing season, determined by a steady transition of the average daily temperature through + 5 °, in the city of Tomsk begins in the first decade of May. The duration of the warm period with an average daily temperature > 0 is 185 days, and the frost-free period is 95 days (86 - 148 days). The risk of frost exists until mid-June and increases again from mid-August (Table 2.2). Weather conditions of the experiments The experiments were conducted from 1986 to 1997. Statistical processing of these data was carried out using correlation, variance and discriminant methods.

The years of observation differed significantly from each other in meteorological conditions. May was warm with frequent precipitation in 1987, 1990 and 1991, cool with frequent precipitation in 1988 - 1989 and dry and cool in 1986. June was characterized by sufficient heat and moisture supply in 1988 -1991 and a shortage of heat and moisture in 1986. lack of heat with sufficient moisture in 1987. July was moderately warm with a lack of precipitation in 1988 and 1989, warm with frequent rains in 1987, 1990 and 1991, and cool and dry in 1986. August and early September turned out to be warm and dry in 1987 and 1991, cool with frequent precipitation in 1988 and 1989, and warm and humid in 1986. In 1992 there was a prolonged and cold spring, a cold June, rainy August. The year was generally unfavourable and atypical for the zone. In 1993, the first and second decades of May were characterized by heavy precipitation. The third decade was favourable for carrying out spring fieldwork. The average temperatures and precipitation in the summer months were close to the long-term average.

From August 25 to September 23, precipitation fell almost daily. The growing season of 1994 was characterized as warm and dry. The average monthly air temperature in the summer months was slightly higher than the long-term average. Precipitation fell in May – 91%, June - August - 77% of the norm. The first autumn frost came on September 3. Overall, the year was atypical for the zone. In 1995, spring started very early. The snow cover came off on April 10. May was cold, windy and dry. Half of the average monthly precipitation fell. The first and second five days of July were cold with an average air temperature of 8.7 and 10.7°. During the same period, a large amount of precipitation fell - 20.5 and 32.8 mm, as a result of which no shoots of thermophilic crops were obtained, and the weather was hot and dry in the future

In 1996, the snow cover came off on April 17. The air temperature in May and June was at the level of the long-term average, while precipitation fell by 214% to the average monthly level. July turned out to be hot and dry. In August, the average monthly air temperature was only 12.6 degrees, and 104.3 mm of precipitation fell (with a norm of 66.3 mm). Lack of heat and excess moisture delayed the ripening of many crops. In 1997, the snow cover came off on April 10, which was 10 days earlier than the previous one. The spring was warm, and the average temperatures in April-May were above the long-term ones. Precipitation in May practically did not differ from the average long-term precipitation. In the summer months, the average temperatures were below the long-term average. Precipitation in August amounted to 114.4 mm with a norm of 67 mm, which significantly affected the yield. In general, the growing season turned out to be cool with an uneven distribution of precipitation. Thus, according to the totality of hydrothermal indicators, 1986 can be considered the most favourable for thermophilic plants, the least in 1989 and 1995. The years 1986, 1991, 1993 and 1995 were favorable for cold-resistant plants.

2.3. Research Methods

2.3.1. Obtaining a Stimulator–Activated Water

The dielectric body was made in the shape of a cube (Pasko, Pat. 2375313 RU). A flat graphite anode and a titanium cathode were used; as a neutral (tarpaulin) diaphragm. Before the experiment, the electrode surfaces were cleaned from the application of salts. During the electrolysis, a constant voltage of 200 V was applied to the electrodes for 15 minutes. Two modifications of ECAW were obtained: anolyte (acidic solution with positive Eh) and catholyte (alkaline solution with negative Eh). Then the treated tubers were planted in the field. Observations of plant development and harvesting were carried out according to generally accepted methods (Dospehov, 1985) (Figure1).

Figure 1. Harvesting potatoes in an experiment. The following are taken into account: the number of tubers, the mass of tubers, the average weight of a tuber from each bush, the degree of disease and pest infestation in different versions of the field experience

2.3.2. Study of Potato Yield and Shelf Life When Processing Tubers With Electrochemically Activated Water

The main accounting unit was the biological yield of Nevsky potatoes on a plot of 57.75 m2, in a shape close to a square of 7.7 x 7.5 m. The layout of all plots was carried out so that there were 11 rows on each plot and 25 equally spaced landing holes in each row. The planting was carried out with pre-selected tubers, which had dimensions of 22 - 27 mm. The fraction of 22-24 mm was soaked in water for 4 hours and planted on experimental plots. The fraction of 25-27 mm was not processed and was used for laying control areas. During harvesting and sorting, all diseased or dried tubers were discarded. A planting set of 270 tubers was prepared for each plot. The planting was carried out manually in pre-prepared ridges along a measuring wire with markers. The distance between the nests was 30 cm, and the planting depth was 10 cm from the top of the ridge.

Control and experimental plots were placed in a specially designated area in a staggered manner, which allowed for each experimental plot to have 3-4 control plots and thus take into account the variability of yield by area. The seed material was processed by completely immersing the tubers in a mixture of anolyte and catholyte. The holding time of tubers in solution was 4 hours. During the treatment, the surface of the water had free contact with the air.l The experienced plots were placed randomly in the area.

The experiments were performed on tubers of the Nevsky variety of super elite qualification, calibrated through a ring with a diameter of 40 mm. The dependence of yield on the composition of the mixture was studied in the entire concentration range — from pure anolyte to pure catholyte. According to the test data, an average yield increase of 20% was obtained with absolute yield values in the control of 250-300 kg/ha. Significant instability of the position of the maximum yield from the concentration of the mixture was noted:

1993 — W = 0 and 0.3;
1994 — W - 0 and 0.9; — W = 0, 0.1 and 1.0;
1999 — W = 0, 0.4 and 1.0.

Therefore, even though the observed intermediate maximum yield on the catholyte/anolyte mixture slightly exceeds the yield increase at W = 0 and 1, in the practical implementation of the method it is more profitable to use pure catholyte and a mixture of catholyte with anolyte (70/30%), providing as a result of their application a stable annual increase in yield at the level of 20%.

Field experience was laid in areas of the Golovnino hospital (Fig. 2) according to the following scheme:

1. - control;
2. - sodium humate;
3. - sodium humate diluted with catholyte (1:15);
4. - catholic;
5. - sodium humate + a mixture of catholyte and anolyte (1:15);

Figure 2. Conditions for conducting an experiment on the effect of processing tubers by various means on their safety in an industrial potato storage facility

The main accounting unit was the biological yield of Nevsky potatoes on a plot of 57.75 m2, in a shape close to a square of 7.7 x 7.5 m. The layout of all plots was carried out so that there were 11 rows on each plot and 25 equally spaced landing holes in each row. The planting was carried out with pre-selected tubers, which had dimensions of 22-27 mm. The fraction of 22-24 mm was soaked in water for 4 hours and planted on experimental plots. The fraction of 25-27 mm was not processed and was used for laying control areas. During harvesting and sorting, all diseased or dried tubers were discarded. A planting set of 270 tubers was prepared for each plot. The planting was carried out manually in pre-prepared ridges along a measuring wire with markers. The distance between the nests was 30 cm, and the planting depth was 10 cm from the top of the ridge.

Control and experimental plots were placed in a specially designated area in a staggered manner, which allowed for each experimental plot to have 3-4 control plots and thus take into account the variability of yield by area. The seed material was processed by completely immersing the tubers in a mixture of anolyte and catholyte. The holding time of tubers in solution was 4 hours. During the treatment, the surface of the water had free contact with the air.| The experienced plots were placed randomly in the area.

The experiments were performed on tubers of the Nevsky variety of super elite qualification, calibrated through a ring with a diameter of 40 mm. The dependence of yield on the composition of the mixture was studied in the entire concentration range — from pure anolyte to pure catholyte. According to the test data, an average yield increase of 20% was obtained with absolute yield values in the control of 250-300 kg/ha. Significant instability of the position of the maximum yield from the concentration of the mixture was noted:

1993 — W = 0 and 0.3;

1994 — W - 0 and 0.9; — W = 0, 0.1 and 1.0;

1999 — W = 0, 0.4 and 1.0.

The authors do not Therefore, even though the observed intermediate maximum yield on the catholyte/anolyte mixture slightly exceeds the yield increase at W = 0 and 1, in the practical implementation of the method it is more profitable to use pure catholyte and a mixture of catholyte with anolyte (70/30%), providing as a result of their application a stable annual increase in yield at the level of 20%.

Field experience was laid in areas of the Golovnino hospital (Fig. 2) according to the following scheme:

1. - control;
2. - sodium humate;
3. - sodium humate diluted with catholyte (1:15);
4. - catholic;
5. - sodium humate + a mixture of catholyte and anolyte (1:15);
6. - a mixture of catholyte and anolyte (1:1) - “mixture”;
7. - sodium humate + anolyte;
8. - anolyte.

2.3.3. Objects of Research

Nevsky variety is a medium–early, table-purpose variety. The tubers are rounded, elongated with a smooth yellowish peel and pink eyes, have a maximum weight of 200 grams, are easy to clean and clean. The white, creamy flesh does not darken for a long time on the cut and does not boil (Figure 3 and Figure 4) The starch content at an average level is no more than 15%. Marketability – 90-95%, yield – 38-50 t/ha. Each tuber brings up to 15 tubers. The mass of the commercial tuber is 90-130 g. Low–growing bushes with a large number of erect stems tend to quickly restore foliage in case of damage by the Colorado potato beetle. Nevsky does not take more than 90 days to fully mature, starting from the time of emergence. The variety is practically not afraid of drought and short-term waterlogging. It is resistant to many diseases. The shelf life is good. The rest period is long. The taste is satisfactory. The value of the variety: resistance to cancer, rhizoctoniosis, alternariasis and blackleg, high yield, excellent environmental.

Figure 3. Appearance of bushes and tubers

Figure 4. Appearance of Nevsky potato variety

Lugovskaya variety is a medium–ripened, table-purpose variety. The tuber is oval, light pink, the skin is smooth, the eyes are small, small, the flesh is white, not darkening when cut (Figure 5 and Figure 6). The eyes are almost invisible. Starch content – 12-19%, yield - 52 t/ha (State export testing). The mass of the commercial tuber is 85-125 g. The value of the variety is: high yield, alignment of tubers, excellent taste qualities, relative resistance to late blight, high storage quality. it does not boil well watery. Maximum yield: 514 kg/ha. Marketability: high. Transportability: not suitable for long-distance transportation. The number of tubers per bush: 10-15 pieces. The shelf life is good. The shrub is erect, of medium height; the leaves are medium-sized, dull dark green in color.

Figure 5. Appearance of bushes and tubers

Figure 6. Appearance of Lugovskaya potato variety

RESEARCH RESULTS

3.1. The Effect of ECAW on Potato Yield

In 1990-1992, experience in processing planting tubers of potato of the Ideal variety was laid at the sites of the Golovnino hospital in the Tomsk region. Its results are presented in Tables 1 - 2.

The effect of treatment of potato tubers with growth and development stimulants on plant productivity

Table 1. The effect of tuberization and spraying of vegetative potato plants on its yield

Option processing	Productivity tubers					Increase in yield compared to control	
	large		small		total		
	kg/m²	%	kg/m²	%	kg/m²	kg/m²	%
Control (Tap water)	2,77±0,21	77	0,84±0,1	23	3,61		
Sodium humate	3,18±0,30	87*	0,48±0,1	13*	3,66	0,05	1
Sodium humate + catholyte	3,18±0,29	84*	0,61±0,2	16*	3,79	0,18	4
Catholyte	4,22±0,3	92*	0,38±0,1	8*	4,70	1,09	22*
Sodium humate + mixture of catholyte and anolyte	3,46±0,27	90*	0,40±0,1	10*	3,86	0,25	7*
Catholyte + anolyte	2,94±0,28	76	0,92±0,2	24	3,86	0,25	7*
Sodium humate + anolyte	2,53±0,14	72	0,99±0,2	28	3,52	-0,09	-2
Anolyte	2,37±0,21	78	0,68±0,1	22	3,05	-0,56	-16*

*Differences at a significance level of 5%.

The maximum increase in total yield was obtained using catholyte (18%). At the same time, the proportion of large tubers increased (9%). The anolyte had an inhibitory effect (a 16% decrease in yield), but the ratio of the proportion of large and small tubers was close to control. Additional treatment of vegetative plants with growth stimulants had a positive effect on yield. The maximum increase in yield was noted when using pure catholyte (30%). Anolyte reduced productivity (by an average of 16%). Activated water was also used to treat potato tubers before storage (for comparison, well-proven zeolite, peat and hydrochloric acid were used. The production experiment was carried out in the potato storage facility of the Tomsk State Agricultural Research Station.

The results of the analysis carried out 6 months after storage revealed the bactericidal effect of the anolyte. The treated tubers also remained dormant for longer and germinated slightly (Table 2).

Table 2. The effect of various methods of processing potato tubers on their safety during winter storage

Option treatment	Proportion of tubers, %, affected by '				Proportion of sprouted tubers, %
	phyto-fluorosis	dry rottenness	wet swarm rotten	total	
Variety Lugovskoy					
Control (without processing)	4	10	14	28	24
Sapropel	4	8	20*	32	8*
Peat	-	10	12	22	12
Zeolite	8	10	6*	24	18
Water	60	2*	4	66*	14
Hydrochloric acid	4	6	2*	12*	20
Anolyte	6	2*	_*	8*	3*
Variety Nevsky					
Control (without processing)	2	3	2	7	16
Sapropel	-	3	-	3*	8*
Peat	-	1	-	1*	7*
Zeolite	-	-	-	-	5*
Water	2	4	3	9	11
Hydrochloric acid	1	1	1	3	4*
Control (without processing)	-	1	1	2*	2*

*Differences at a significance level of 5%.

3.2.The Effect of Pre-Planting Treatment of Seed Potatoes With Thermally and Electrochemically Activated Water on Its Yield

In the conditions of the West Siberian region, potato seed material has been in storage for almost eight months. During storage, tubers lose moisture, therefore, at the stage of pre-planting preparation of potatoes, it is necessary to replenish the content in tubers to normal and ensure correlation of biochemical parameters associated with plant growth and development in subsequent phases. The simplest and most affordable ways to solve these problems may be heat treatment of seed tubers with water at a given temperature and/or treatment with electrochemically activated water. Deep warming of potato seed tubers after the end of natural rest activates enzymatic reactions associated with the transformation of stored substances in tubers necessary for the development of growth processes, and replenishes the necessary lag.

Below are the results of field experiments on the effect of pre-sowing treatment of potato seed tubers with thermally and electrochemically activated water on its yields. According to the results of a field experiment, it was found that processing Nevsky potatoes with a mixture of catholyte and anolyte leads to an increase in yield. A beneficial effect on yield was noted, but no definite pattern was established from the presented data. In this regard, subsequent experiments were conducted in the direction of more careful selection of seed material and expansion of the concentration composition of the used mixtures of catholyte and anolyte. The experiments confirmed the conclusion about an increase in potato yield when soaking the planting material in electrochemically activated water (table 3).

Table 3. Results of a field experiment on the yield of Nevsky potato varieties during pre-planting treatment of tubers with a mixture of catholyte and anolyte

Content anolyte in mixture, %	Given productivity on the experimental plot, c/ha	Average yield on the test plots, c/ha	Increase yield, %	Average increase yield, %
0	248	204,5	21,3	23,6
	274*	217,5*	26,0*	
10	228	231,0	-1,3	9,4
	277*	230,5	20,2	
20	224	204,0	9,8	16,6
	298*	241,7*	23,3*	
30	256	216,0	18,5	28,3
	312*	225,0*	38,6*	
40	236	221,0	6,8	9,6
	262*	233,0*	12,4*	
50	244	216,0	13,0	16,0
	266*	222,0*	19,0*	

*Differences at a significance level of 5%.

In parallel with experiments to study the effect of electrochemical water on potato yields, pre-sowing soaking of potato tubers was carried out for four hours in water (tap or taken from natural sources) of different temperatures. The results of the experiments are presented in Table 10, from which it can be seen that there is a monotonous increase in potato yields of various varieties as the water temperature increases. Heat treatment modes up to 40 ° C are considered safe for seed material (table 4).

Table 4. The effect of water temperature during pre-sowing treatment of tubers on the yield of various potato varieties

Variety potatoes	Average yield, g/bush, at water temperature, °C				Average yield in control, g/bush	Yield increase at 40°C, %
	35	40	45	50		
Nevsky	650	680*	750*	920*	600	13
Spring	400	500*	650*	720*	350	42
Nikulinsky	900	930*	940*	1010*	830	12
Santa	600	620*	640*	650*	570	9

*Differences at a significance level of 5%.

The greatest effect was manifested as a result of combined exposure, when, after preliminary exposure in water, seed potatoes were washed for 15 minutes in water with a temperature of approximately 50 ° C. This regime, apparently, is aDWisable to use for the improvement of seed material. It should be noted that the introduction of these methods of preparation makes it possible to block the main channel of pathogenic microorganisms returning to the soil with seed material. Thus, according to the results of field experiments, it is shown that the use of pre-planting treatments of seed material with electrochemically activated water in the form of anolyte and catholyte and ordinary water at a temperature of 40 ° C in potato growing practice is a reliable agricultural method for increasing potato yields.

CONCLUSION

Activated water is not only an effective but also a cheap stimulator of plant growth and development. This is especially important when introducing technology into the manufacturing sector. In addition to its high efficiency, it preserves the ecological purity of the soil and agricultural products, since in the relaxation process, the properties of water return to the initial level, and no additional substances are introduced into the environment.

In our large-scale production experiments, the high efficiency of using electrochemically activated water has been shown to increase the yield of potatoes and improve the safety of potatoes during winter storage by reducing the incidence of tubers and their germination. In all experiments, an increase in the overall yield was noted, and, most importantly, the early, most profitable yield of vegetable products.

During the experiments, three-year field tests were carried out before planting the club with external modifications of electronically activated water in the conditions of risky farming of economic chemistry in Siberia. The maximum increase in the total tuber yield was obtained using catholyte (18%). As a result, the proportion of large tubers (9%) at harvest increased. Anolyte had an inhibitory effect; it reduced the yield by 16%), but the proportion of large and small tubers as a result of yield was close to the control.

Additional treatment of vegetative potato plants with plant growth and development stimulants increased the yield. The maximum increase in yield was observed when using pure catholyte (30%). Anolyte reduced yield by an average of 16%. Potato tubers were also treated with activated water before storage. For comparison, we used well-proven zeolite, peat and hydrochloric acid. The production experiment was carried out in a large-scale experiment in the potato storage facility of the Tomsk State Agricultural Research Station. The results of the analysis carried out 6 months after storage revealed the bactericidal effect of anolyte. The treated tubers also remained dormant longer and germinated poorly. In the variant with the Lugovskoy variety, they were 8% less affected by dry rottenness and were not affected by wet swarm rotten at all (in the control – 14%).

The overall incidence of potato tubers during winter storage was 20% lower than in the control (without treatment) and almost 50% lower than when treated with water. Treatment of tubers before storage with sapropel, zeolite and hydrochloric acid showed a fungicidal effect, but to a much lesser extent. The potato variety Nevsky in the control turned out to be resistant to damage by pathogenic microflora, so the effectiveness of treating tubers before storing for winter storage turned out to be much less pronounced (5%). The experiments confirmed the conclusion about an increase in potato yield when soaking the planting material in electrochemically activated water A study of the influence of ordinary water temperature on potato yields revealed a monotonous increase in potato yields of various varieties as the water temperature rises to 40 °C, which is considered safe for seed material. The greatest effect was manifested as a result of combined exposure, when, after preliminary exposure in water, seed potatoes were washed for 15 minutes in water with a temperature of approximately 50 ° C. This regime is advisable to use for the improvement of seed material. It should be noted that the introduction of these methods of preparation makes it possible to block the main channel of pathogenic microorganisms returning to the soil with seed material. Thus, according to the results of field experiments, it is shown that the use of pre-planting treatments of seed material with electrochemically activated water in the form of anolyte and catholyte and ordinary water at a temperature of 40 ° C in potato growing practice is a reliable agricultural method for increasing potato yields.

It should be noted that despite the high efficiency of using activated water in potato farming, it is rarely used. This is due to the low productivity of the results, and the desire to get high profits without proper methodological preparation. Our experience shows the need for the strictest consideration of the composition of the source water, its treatment regime, and control of physicochemical properties after treatment, on the one hand, taking into account the characteristics of the biological, physiological, environmental properties of the treatment object (properties of the variety, condition of tubers, etc.) and the conditions of the experiment (in the field - soil, hydrothermal regime of the growing season; in storage - temperature, humidity, substrate). Currently, this area is at the stage of collecting and accumulating factual material, so the number of publications is insignificant.

At the stage of analysis and modelling, the tasks we have indicated will undoubtedly be solved, and the use of activated water will take its rightful place among effective, environmentally friendly and cheap technologies.

The prospects for further research are related to the study of the mechanism of the effect of activated water on the germination of tubers for their stimulation (before planting) and inhibition (before laying for winter storage), as well as the possibility of developing industrial electrolyzers for large-scale use. To achieve good reproducibility of experiments, it is necessary to take into account a set of indicators of activated water, such as pH, redox potential, hardness, chlorine content, various forms of oxygen, calcium, magnesium, etc. It is necessary to build mathematical models and study the physiological mechanisms of stimulation and inhibition of the main metabolic processes triggered by various modifications of activated water.

REFERENCES

Afek, U., Orenstein, J., & Kim, J. J. (2001). Control of silver scurf disease in stored potato by using hydrogen peroxide plus (HPP). *Crop Protection (Guildford, Surrey)*, *20*(1), 69–71. DOI: 10.1016/S0261-2194(00)00054-5

Anwar, A., Xianchang, Yu., & Jansu, L. I. (2020). Soaking seeds as a promising method to improve growth, chlorophyll, photosynthesis and nutrient content in cucumber seedlings. *Notulae Botanicae Horti Agrobotanici Cluj-Napoca*, *48*(1), 116–127. DOI: 10.15835/nbha48111806

Bradford, KJ, Steiner, JJ, Travata, S.E. (1990). Analysis of seed seedling for germination and appearance of batches of pepper seeds. *The science of crop production,* 30, 718–721.

Damalas, C., Koutroubas, S. D., & Fotiadis, S. (2019). The effect of the hydroprimer on seed germination and field productivity of horse beans during spring sowing. *Agriculture*, *9*(9), 201. DOI: 10.3390/agriculture9090201

Devika O.S., Singh S., Sarkar D., Barnwal P., Suman J. and Rakshit A. (2021). Seed seeding: a potential complement to integrated resource management in fragile intensive ecosystems. *The boundaries of sustainable food systems*, 5. .DOI: 10.3389/fsufs.2021.654001

Giri, G.S., Schillinger, V.F. (2003). Seedling of winter wheat for germination, germination and yield. *The science of crop production,* 43, 2135-2141.

Kefeli, V.I. (1978). Plant growth and natural regulators. *Plant physiology,* 25, 5, Moscow, Science. Al-Mughrabi, K. I. (2007). Effect of treatment of potatoes in storage and pre-planting with hydrogen peroxide (H2O2) on emergence and yield.

Kovalenko, L. V., & Folmanis, G. E. (2006). Biologically active iron nanopowders. *Science*, •••, 124.

Lugovskoy potatoes: subtleties, yield, cultivation and care URL: https://dzen.ru/a/YlVl2BQqfmgppBSK. Date of acessed 27/08/2024

Norikane, J. H., Brook, R. C., & Kirk, W. W. (2001). Efficacy of purogene and oxidate disinfectants added to potato storage humidity water for pathogen control. *American Journal of Potato Research*, *78*, 473–474.

Olsen, N., Kleinkopf, G., Secor, G., Woodell, L., & Nolte, P. (2000). The use of chlorine dioxide in potato storage. *Bulletin 825,* University of Idaho, 4.

Pasko, O.A. (1991). Acceleration of seed germination when treated with electrochemically activated water. *Physiology and biochemistry of cultivated plants*, 23 (6), 552 – 556.

Pasko, O. A. (2000). *Activated water and its use in agriculture: monograph*. Tomsk: TPU, 180.

Pasko, O. A. (2000a). *Ecological aspects of increasing the productivity of flower, vegetable crops and potatoes in the taiga zone of Western Siberia*: abstract of the dissertation... Doctor of Agricultural Sciences: specialty 06.01.09, 03.00.16. Novosibirsk State Agrarian University. Univ.: Novosibirsk, 32.

Pasko, O. A. (2009). Pat. 2375313 RU, IPC C 02 F 1/461. *Flow-through diaphragm electrolyzer* No. 2007118618/15. Published on 10.12.2009.

Pasko, O.A. (2010a). The growth and development of plants stimulated by electrochemically activated water. *Bulletin of the B.S.S.A. named after V.R. Filippov scientific and theoretical journal*, 3, 54-59.

Pasko, O. A. (2010b). Stimulation of vegetative reproduction of ornamental plants using electrochemically activated water. *Phytodesign in modern conditions materials of the International Scientific and Practical Conference*, Belgorod: BelSU Publishing House, 146-147.

Pasko, O. A. (2010c). Physico-chemical changes in tap water during its treatment in various ways. Property changes of piped water treated with different methods. *Water: Chemistry and Ecology*, *7*, 40–45.

Pasko, O. A. (2011). The influence of pre-sowing stimulation of cucumber seeds on yield. *Agricultural Sciences*, *8*, 20–22.

Pasko, O.A. (2011a). Stimulation of tomato seeds with activated water increases the yield. *Potatoes and vegetables*, 6, 21.

Pasko, O. A. (2011b). Stimulation of vegetative reproduction of plants with sodium humate Innovative aspects of peat extraction, processing and application. *Proceedings of the International Conference dedicated to the 115th anniversary of the National Research* Tomsk Polytechnic University, Tomsk: TPU Publishing House, 122-125.

Pasko, O. A. (2011b). The effect of presowing stimulation of cucumber seeds on yield. Influence of presowing stimulation of cucumber seeds on productivity. Agricultural science. Scientific. *Theoretical and Industrial Journal*, *8*, 20–22.

Pasko, O. A. (2024). *Method of obtaining stimulator of growth.* Retrieved from URL https://patents.google.com/patent/SU1574196A1/en Date of acessed 27/08/2024

Pasko, O. A., & Gomboev, D. D. (2011). *Activated water and the possibilities of its application in crop production and animal husbandry*. Tomsk Polytechnic University Publishing House.

Pasko, O. A., & Ketskalo, V. M. (1991). The influence of electrochemically activated water on seed germination. *Plant Physiology*, *38*(2), 381–385.

Pasko, O.A. & Leonov V. P. (2011) The mechanism of action of activated water on rhizogenesis Cherenkov. The mechanism of activated water on rhizogenesis cuts. *Modern problems of science and education*, 5, 10.

Pasko, O.A. & Semenov, A.V. (1991). *Method of pre-sowing seed treatment.* Copyright certificate No. 1667674 of the USSR, class A 01 C 7 /00. Publ. 07.06.1991. Byul. No. 29.

Pasko, O.A., Semenov, A.V. (1993a), *A method for obtaining a plant growth and development stimulant.* Copyright certificate No. 161 19557 USSR, class A 01 N 59/00. Publ. 06/07/1993. Bul. No. 21.

Pasko, O.A., Semenov, A.V. (1993b) *A method for obtaining a plant growth and development stimulant.* Copyright certificate No. 16119557 of the USSR, class A 01 N 59/00.

Pasko, O. A., Semenov, A. V., Smirnov, G. V., & Smirnov, D. G. (2007). *Activated liquids, electromagnetic fields and flicker noise. Their application in medicine and agriculture.* TUSUR Publishing House.

Potato variety Lugovskoy: characteristics and description, reviews, photos. URL: https://fermilon.ru/sad-i-ogorod/ovoshhi/sort-kartofelya-lugovskoy-harakteristika-i-opisanie-otzyvy-foto.html. Date of access 08/27/24.

Potato variety Nevsky: features, planting and care. URL: https://ferma.expert/rasteniya/ovoshchi/kartofel/kartofel-nevskij. Date of access 08/27/24.

Potato variety Nevsky. State variety testing. URL: https://gossortrf.ru/registry/gosudarstvennyy-reestr-selektsionnykh-dostizheniy-dopushchennykh-k-ispolzovaniyu-tom-1-sorta-rasteni/nevskiy-kartofel. Date of access 08/27/24.

Secor, G. A., & Gudmestad, N. C. (1999). Managing fungal diseases of potato. *Canadian Journal of Plant Pathology*, *21*(3), 213–221. DOI: 10.1080/07060669909501184

Semenenko, S. Ya., Belitskaya, M. N., & Likholetov, S. M. (2013). Phytosanitary improvement of grain and vegetable crops using electrochemically activated water // Advances in modern natural science. 1, 78-82. URL: https://natural-sciences.ru/ru/article/view?id=31286 (date of access: 08/27/2024).

Tasty and productive Lugovskoy potatoes. URL: https://rusfermer.net/ogorod/korneplody/kartofel/sorta/srednerannie/lugovskoj.html. Date of access 08/27/24.

Chapter 6

The Influence of Pre-Sowing Treatment of Vegetable Seeds With Activated Water on Yield

Olga Anatolievna Pasko
https://orcid.org/0000-0001-8322-834X

National Open Institute St. Petersburg, Russia & Leningrad Branch of the Federal State Budgetary Institution Agrochemical Service of Russia, St. Petersburg, Russia

Nadezhda Anatolievna Lebedeva
https://orcid.org/0000-0003-4095-2631

International Personnel Academy, Germany

ABSTRACT

The chapter presents the main approaches to the introduction of environmentally friendly stimulants for the growth and development of cultivated plants into production. The purpose of the study. Identification of opportunities to increase the yield of vegetable crops through the use of effective, environmentally friendly and affordable growth and development stimulants. The object of the study is vegetable crops (cucumbers and tomatoes) grown in open and protected ground. The chapter describes the successful experience of industrial testing of water acti-vated by various

DOI: 10.4018/979-8-3693-4392-0.ch006

methods. The high efficiency of pre-sowing seed treatment with a mixture of catho-lyte and anolyte is shown. The yield of tomatoes in protected soil increased by 44% and cucumber by 33%, while the quality of the products met the appropriate requirements.

INTRODUCTION

The priority task of the current stage of any country's development is to ensure its food security based on increasing crop yields and improving the quality of products. The real reserves for achieving this task are modern phytotechnical techniques that accelerate the growth and development of plants and increase yields. Their application in the conditions of the taiga zone of Western Siberia is complicated by the sharp variability of the meteorological conditions of the growing season. This makes it necessary to study the influence of hydrothermal factors on plant growth and development, identify critical periods in their lives, and develop ways to adequately correct plant growth and development through effective and environmentally friendly stimulants.

The Purpose of the Study

Identification of opportunities to increase the yield of vegetable crops by using effective, environmentally friendly and affordable growth and development stimulants; improvement of floriculture based on the identification of promising species and varieties of annual flowering plants, studying the biology of their development, increasing productivity and improving seed production. The object of the study is vegetable crops (cucumber and tomato) grown in open and protected ground. The subject of the study is the possibility of increasing yields and improving product quality through the pre-sowing treatment of seeds/tubers with environmentally friendly growth and development stimulants.

Practical Significance and Implementation of Research Results

Recommendations based on scientific provisions are possible for state, peasant (farmer) agricultural enterprises, and members of horticultural cooperatives.

BACKGROUND

One of the most important areas of modern crop production is the development of effective phytotechnical techniques to increase the yield of cultivated plants and reduce their dependence on weather conditions. For this purpose, factors of influence of physical and chemical nature are widely used in practice (Basra, Farook et al., 2005; Javed et al., 2020; Sivasubramaniam et al., 2011; Benazzuk, Jazuli et al., 2019). Their effectiveness largely depends on the season, geographical, soil-climatic and several other factors, as well as compliance with agricultural techniques and experimental methods. At the same time, there are many similarities in the action of various factors.

1.1. Characteristics of Plant Growth and Development Stimulants

The factors of physical impact used in crop production include electromagnetic radiation, electric field (constant, alternating, high frequency, corona discharge), magnetic field, ionizing radiation - alpha, beta, gamma rays, neutrons (fast, intermediate, slow), protons, other high-energy corpuscular radiation, as well as ultrasound (Shera, Scholz, 2021; Pandiselvam, Mayukha et al, 2020).

The action of chemical stimulants is fundamentally different from the action of physical ones, which bring energy, not substance, into plant tissues (Mustafa et al., 2022). On the other hand, unlike fertilizers, they are not nutrients (Samarah et al., 2020). Chemical stimulants are factors in controlling the growth and development of plants. They are divided into growth-regulating substances and mutagens (Afzal et al., 2020). Among growth-regulating substances, regulators, herbicides, retardants and natural stimulants of a complex chemical nature are distinguished (Kadena, Preston et al, 2018; Hinojosa-Davalos, Arias-Rios et al., 2020). Note that the positive effect of stimulants is manifested only on a high agronomic background, assuming a sufficient amount of nutrients, moisture and light.

It has been established that it has a hormonal effect on plants, improves the absorption of nutrients by the root system, and is actively involved in redox processes. Increases the resistance of plants to diseases is low-toxic. Let's note one property common to all stimulants. They contribute to a sharp increase in plant growth. Therefore, the use of stimulants should be accompanied by enhanced nutrition and moisture supply. It should be emphasized that the morphological and physiological manifestations of pre-sowing seed treatment with different means are identical (Stadler,1928; Sakala, Demchuk et al.,2012; A Pasko & Ketskalo, 1991; Rybinsky, 2001) (Table 1).

Table 1. Reactions of plant objects to the action of various physical and chemical agents according to a number of generalized biological indicators

Biological indicator	Electromagnetic radiation				Ionizing radiation			Electromagnetic fields					Ultra-sound	Regulators	Hb	NS	Mgn
	γ*, R	UR	IR	L	η	ρ	e	I =	I≈	HFC	KD	MF					
Activation of growth processes (rate of cell division, accumulation of dry matter)	+	+	+	+	+	+	+	+	+	+	+	+	+	+	+	+	+
Changes in the physicochemical state of the cell (permeability, viscosity, pH)	+	+	+	+	+		+		+	+	+		+	+			+
Increased photosynthesis	+	+	+	+		+	+		+		+	+	+	+	+	+	+
Activation of metabolism (proteins, carbohydrates, hormones)	+	+	+	+		+	+		+		+		+		+	+	+
Possibility of modifying the reaction by other factors	+	+	+			+	+	+			+	+	+		+	+	+
The appearance of morphoses	+	+		+	+		+	+		+	+	+		+		+	+
The appearance of free radicals	+		+	+		+	+		+	+		+	+	+	+	+	+

*Note: γ – y-radiation, R – X-ray radiation, UR – ultraviolet radiation, IR – infrared radiation, L –laser radiation, η – neutrons, ρ – protons, – accelerated electrons. I = direct electric current. I≈ alternating electric current, HF– high-frequency current, KD – corona discharge, MF – magnetic field, Hb – Herbicides; NS – Natural stimulants; Mtg –Mutagenes

Increased activity of physiological processes under the influence of growth and development regulators in the optimal conditions for the plant species and variety promotes acceleration of plant growth and development (Tretyakov, 2005) and increases yields These minimize production costs and help reduce production costs, which means they lead to increased economic benefits. From an environmental point of view, it is important to improve the quality of vegetable products, in particular, to reduce the nitrate content (Komarov, 2005; Komarov, 2004).

1.2. Efficiency of Application of Influencing Factors in Crop Production

The factors of physical nature have found wide application in crop production, since their use in optimal doses has a wide multilateral effect on crops, causing positive changes in their growth, development and yield. For example, pre-sowing irradiation of winter wheat Bezostaya-1 X seeds increased their germination energy by 5-7%, productive business by 7%, tree content by 0.6-1.3%, and gluten by 3-4%. Grain yield increased by 0.4-0.6 t/ha. Irradiation of spring wheat seeds caused similar shifts: germination energy increased by 8-11%, protein content - by 0.2-1.1%, and gluten - by 2.0-3.4%.Grain yield increased by 0.2-0.4 t/ha. Note that in that and subsequent versions, ICSS with a capacity of 44.7 cal/cm2 was used for 45-360 minutes (i.e., the total irradiation was 7.4-380.4 cal/ cm2.

Six-year tests on the use of ICSS on table beet seeds, conducted in Kazakhstan, revealed a stable stimulating effect. It manifested itself in an increase in the yield of root crops by an average of 7.0 t/ha (or 15.2%) and an increase in the sugar content by 0.7-0.9% (Recommendations ..., 1976). Table beet seeds turned out to be sensitive to the effects of laser irradiation, although the actual effect of its use was somewhat lower. The increase in the yield of root crops averaged 56.2 c / ha, with an increase in the sugar content in them - 0.4-0.5%. In this case, the LG-75 helium-neon laser was used for -180 min. In the same series, the effect of the pre-sowing treatment of table beet seeds with a corona discharge field on its yield qualities was studied. A corona discharge with a field strength of 5.6-.0 V/cm with exposures of 3-60 seconds was used. An increase in yield was established by 6-7%, and sugar content - by 0.4-0.6%.

Irradiation of cucumber and tomato seeds in the previously indicated modes increased their soil germination by 11-13%. The early yield of cucumber in the open ground increased by 28%, and the total - by 22%. For tomatoes, these indicators look somewhat different - the increase in tomato yield was: in the open ground early - 30%. total - 18-20%; under a tunnel-type film shelter - 6% and 30%, respectively. Consequently, favourable temperature conditions enhance the effect of the use of ICSS. Let's note one more important point. Artificial infection of tomato seed material with black bacterial spotting manifests itself differently depending on its further treatment-1. Thus, light pulse irradiation had a positive effect on the growth and development of vegetative and reproductive organs. This contributed to an increase in the number of healthy tomato plants by 20% and reduced leaf infestation with black bacterial spotting from 1.7 to 1.2 points. An improvement in the commercial qualities of the fruits was noted.

Root crops are also quite responsive to pre-sowing seed training. The increase in the yield of carrot root crops was 34-56% with an increase in carotene content by 11-51%. For radishes, the increase reached 11-33%, and the vitamin C content increased from 17.8% (in the control) to 21.8%. The effect of gamma radiation on cabbage has confirmed its high efficiency. The yield increase was in the range of 17-21%. A higher content of ascorbic acid was found in the heads grown from irradiated seeds.

Production tests conducted on fruit and vegetable crops showed the following: an increase in the yield of cucumber under optimal conditions was 21-40%, tomato - 14-27%, melon, watermelon and pumpkin - up to 14%. An improvement in the quality of marketable products was noted: an increase in the content of vitamin C (19-30%), Sugars (8-36%), melon, watermelon and pumpkin carotenes. For the treatment of vegetable seeds, it is recommended to soak tomato seeds within 24 hours in a solution of methylene blue (0.3 g/l of water), cucumber - in a solution of succinic acid (17 mg / l of water), whether methylene blue (0.3 g / l of water) or baking soda (10 g / l of water); table beet - in a 0.1% solution of boric acid or .2% solution of manganese sulfate.

Studies of the effectiveness of pre-sowing stimulation of seeds with electrochemically activated water have shown significant stimulation of growth and development in twenty species of summer beetles (Pasko, 1991, 2010a, b. c; 2011, 2011d; Pasko & Ketskalo, 1991). The given literature data indicate the multilateral influence of growth stimulants on the physiological processes of plants. With their help, it is possible to accelerate development (the onset of fruiting), increase yields, improve product quality, and facilitate the work of obtaining it. It should be noted the increased resistance of plants to adverse environmental factors of biotic (weather conditions, agrophone) and biotic (diseases and pests) origin. At the same time, the scale of the introduction of stimulants does not yet meet the needs of crop production (proteins, carbohydrates, hormones).

1.3. The Effectiveness of Using Various Modifications of Activated Water as a Stimulator of Growth, Development and Productivity of Agricultural Plants

The effectiveness of the use of various modifications of activated water in agriculture in recent years, along with traditional means of increasing yields (crop rotation, selection of varieties, agrotechnical, protective measures), various methods of processing planting material have been increasingly used. Several biologically active drugs (gidrel, tur, fospinol) have been developed, with the help of which it is possible to target plants to mobilize their biopotential. Managing the growth and development of plants using biologically active substances allows for a significant increase in yield with minimal labour and money. At the same time, the problem of choosing not only effective but also environmentally friendly stimulants, which, in particular, include sodium humate obtained from brown coals and activated water, becomes urgent.

Because the action of factors of the most diverse nature is effective (i.e., to obtain the same result, you can always choose a certain dose of any factor), the criteria of environmental safety, cheapness and accessibility of the stimulant come to the fore (Pasko, Pat. 2375313 RU; Pasko, Semenov, Copyright certificate No. 161 19557; No. 16119557). In our opinion, the so-called activated water meets the complex of these requirements. In recent years, it has become widespread in agriculture. However, its practical application has always been ahead of theoretical research. Therefore, it is not uncommon for the existence of abnormal, biologically active properties of water to be questioned at all due to insufficient scientific evidence.

This causes increased interest among scientists of different specialities in studying the properties and practical application of the differences of its various modifications to accelerate technological processes in biology, agriculture, biotechnology and chemistry, as well as in medicine - for the prevention and treatment of certain diseases.

Magnetized water, due to the simplicity and cheapness of its production, has found wide application in various sectors of the national economy, including agriculture. The "magnetization" of water is carried out by influencing a water flow moving at a speed of 1-5 m / s with a transverse magnetic field with a strength of at least 300 E. The use of magnetized water makes it possible to intensify the process of agricultural production without special additional costs (Vasiliev et al., 2010; Talanova, 2006).

The increase in crop yields after treating seeds with magnetized water was (in per cent): cucumber – 17-37; radishes – up to 5.0; tomato – 10 - 45; peas – 9 - 30; corn – 17 – 45; wheat – 14 - 45; soybean – 10 – 45; rice – 18; sugar beets – 8-9; alfalfa – 2-9; spring barley – 6; cabbage – 11; annual grass – 8%; Sudanese grass – 11; barley – 8; perennial grasses: 1st cutting – 14, 2nd cutting – 10; 3rd cutting – 10; potato – 9.

Currently, extensive material has been accumulated indicating that plants treated with magnetized water develop better, bloom and mature earlier, and are more resistant to adverse factors. The effect of magnetized water is manifested not only in increasing the volume of agricultural products but also in improving their quality. Thus, the fruits of tomatoes from experimental plants had a higher (by 35 g) weight compared to the control ones, contained more sugar - by 0.4% and ascorbic acid - by 1.5 mg /100 g, were characterized by a low nitrate content and had a low acidity. When watering watermelons with magnetized water, the fruits of the experimental variant accumulated a greater amount of dry substances by 7.1%, monosaccharides (fructose - by 7.1%, glucose - by 13.0%), ascorbic acid - by 32.7%.

According to the laboratory of the Institute of Irrigated Agriculture, the gluten content in the wheat grain of the Yershovskaya-32 variety increased by 9.8%, which in turn increased the yield of bread [3,100 kg of flour by 24 kg. Valuable biochemical shifts are also confirmed by foreign researchers. It is interesting that when magnetized water is exposed to the spores of an obligate parasite that causes infusion due to the powdery mildew of melons, spore germination decreases by 27 times. The inhibitory effect of magnetized water on the development of fungal spores causing fusarium has also been noted (cited by Pasko, and Gomboev, 2011).

Pre-sowing soaking of seeds in magnetized water is especially effective in cases where seed germination is difficult due to poor water permeability of the seed holster. Thus, the ground germination of Galega orientalis L. seeds increased by 30-40% after three hours of treatment in magnetized water. Magnetization also had a beneficial effect on slightly mineralized water. The increase in the yield of winter wheat when irrigated with magnetized slightly mineralized water was 14%. It should be noted that the use of magnetized water cannot make up for deficiencies in agricultural technology. The higher the level of agricultural technology, the more effective is irrigation with magnetized water.

Degassed water (DW) is used for the pre-sowing soaking of seeds, and processing cuttings before planting and watering plants. Experiments have shown that degassed water stimulates the growth processes of corn, increases the productivity of fruit trees, the yield of winter wheat - by 25%, spring wheat - by 28%, cucumber - by 50%, cotton - by 10-16%), sugar beet – by 42-48%. Under the influence of the pre-sowing treatment of cotton seeds, the fiber becomes stronger; the size of wheat grains increases, and the number of ears increases (Zelepukhin, Pasko et al., 1988).

The accelerated development of tomato plants contributed to earlier and more harmonious fruit ripening, which began 3 days earlier than in the control. The first fees were higher -6-28%. Soaking of seeds and double spraying of seedlings increased tomato yield by 1.6 kg/m2. The yield of dry matter with the yield of stimulated plants was 14-16% higher than the control. The quality of the fruits improved: the sugar content increased (by 7-13%), vitamin C (by 13-15%), the nitrate content decreased (by 30-42%). The average fruit weight in experimental plants was 72-76 g, and in control plants 69-71 g. The number of non-standard fruits decreased due to a weaker lesion with late blight (Zelepukhin, Pasko et al., 1988).

The effectiveness of using DW on vegetable plants was confirmed at the Department of Vegetable Growing of the Belarusian Agricultural Academy. In a spring film greenhouse on a cucumber of the TSKHA-

211 variety, an average of three years (1980-82) was obtained: 10.6 kg/m in the control, and 13.9 kg/ m in the DW. In the open ground, the yield of cucumber of the Nezhinsky variety was 157 hectares in the control (201 kg/ha for DW), Dolzhik varieties (on average for three years) in the control 266 kg/ha (360.5 kg/ha for DW), Elegant varieties - 213 and 300 kg/ha, respectively. In vegetable growing, activated water is used as an environmentally friendly growth stimulant for pre-sowing seed treatment, growing seedlings and processing plants in greenhouses. For example, the cultivation of tomato and cucumber seedlings took place in the seedling department of greenhouses (Zelepukhin, Pasko et al., 1988).

Pre-sowing seed treatment included: thermal heating (3-4 hours at a temperature of 60-70 °), decontamination (1-2 days), etching in 0.5% potassium permanganate solution prepared with degassed water), thorough rinsing of the seeds of the DW, drying and soaking in the DW for two days before pecking). The seeds were sown in the seedling compartment in peat-humus pots. Seedlings were sprayed twice three times (with an interval of 7-10 days, the first in the phase of two or three real leaves). This treatment ensured earlier and external emergence of seedlings and better seedling growth.

Pre-sowing seed treatment and triple spraying of tomato seedlings gave the following results: the average plant size of a 30-day seedling in the experiment was 18.8 cm high (5.3 cm in the control), leaf area 187 cm (119 cm in the control), seedling weight 10.3 g (6.3 g in the control), root weight 0.82 g (in the control 0.62 g). The growing of seedlings was accelerated by 7-10 days. By the time the seedlings were planted, the leaf surface area of the experimental plants exceeded the control by 37-40%, by the beginning of fruiting -1-28%, and during the period of mass fruiting - 21-32%, remaining at this level until the end of the harvest. Stimulation of DW plants led to an increase in the number of flowers by 14-16%, and ovaries - by 20%. Accelerated plant development led to earlier and more harmonious fruit ripening, which began three days earlier than in the control. The first fees were higher (by 2.6-28%). Soaking of seeds and double spraying of seedlings increased tomato yield by 1.6 kg/m2. The yield of dry matter with the yield of stimulated plants was 14-16% higher than the control.

Production tests of pre-sowing treatment of *electrochemically activated water* (ECAW) agricultural seeds were carried out in 17 farms of Kyrgyzstan, Uzbekistan, Ukraine, Kazakhstan and Russia (Volgograd, Leningrad, Kuibyshev, Tomsk regions, Stavropol, Krasnodar Territory and Tatarstan). The following crops were selected as the studied crops: corn, buckwheat, millet, sunflower, oats, wheat, barley, sweet pepper, tomato, cucumber, radish, lettuce, onion and cabbage. We consider such a growth stimulant as activated water to be the most environmentally friendly. Various assumptions are made about the causes of its abnormal properties (they will be discussed later), but chemically activated water is water with a number of altered physico-chemical properties. After an incident of a certain time (relaxation period) activated water goes into an equilibrium (initial) state and does not pose any danger from an environmental point of view.

2. OBJECTS AND METHODS OF RESEARCH

2.1. Geographical Location and Weather Conditions of the Tomsk Region

The Tomsk Region lies almost entirely in the southeastern part of the West Siberian Lowland, in a zone of risky agriculture, bordering the Kemerovo, Novosibirsk, Omsk, Tyumen regions and the Krasnoyarsk Territory. The elevations of the lowland do not exceed 200 m above sea level. The geographical boundaries of the region are as follows: from south to north - from 55.7 to 61.0° north latitude (600 km), from west and east - from 75.0° to 89.4 ° east longitude from Greenwich (850 km). The territory of the region is 314.4 thousand km2 The taiga zone occupies huge areas - from the polar route to about 57 ° C., i.e. about 60% of the area of Western Siberia. It is heterogeneous in nature. In the north, rare-coniferous spruce-deciduous swampy forests turn into deciduous cedar mixed with pine riverine forests, which occupy most of the zone. The Mesopotamians are swampy and practically treeless.

Further south is a zone of birch-pine-dark coniferous swampy forests. A narrow strip of small-leaved forests (birch, aspen with an admixture of pine) adjoins the forest-steppe zone. The soil cover of this subdistrict is represented by grey forest soils of all three subtypes, as well as sod-podzolic). Leached chernozems are less common. The sum of soil temperatures at a depth of 10 cm for a period with temperatures above 10 ° for clay and heavy loamy mechanical composition is 1600-1800 °, loamy - 1700-1900 °, sandy loam and sandy - 1800-2000 °. Swamp soils are gaining temperatures in the range of 1200 - 1400 °.

The experimental site of the Siberian Botanical Garden is located on grey forest medium loamy soils with humus content in the range of 3-4%, mobile forms of phosphorus and potassium 5-10 mg/100 g of soil. They are characterized by low water and air permeability, and a tendency to swim. The reaction of the soil solution is slightly acidic, it is 5.9-6.0.

2.2. Meteorological Conditions

According to the Agrometeorological Bulletin, the average annual air temperature in the experimental area is -0.6 ° C. The number of days in a year with temperatures above 5 ° is 153, above 10° is 113, and the sum of temperatures above 10° reaches 1750 °. The growing season, determined by a steady transition of the average daily temperature through + 5 °, in the city of Tomsk begins in the first decade of May. The duration of the warm period with an average daily temperature > 0 is 185 days, and the frost-free period is 95 days (86 - 148 days). The risk of frost exists until mid-June and increases again from mid-August (Table 2.2). Weather conditions of the experiments The experiments were conducted from 1986 to 1997. Statistical processing of these data was carried out using correlation, variance and discriminant methods.

The years of observation differed significantly from each other in meteorological conditions. May was warm with frequent precipitation in 1987, 1990 and 1991, cool with frequent precipitation in 1988 - 1989 and dry and cool in 1986. June was characterized by sufficient heat and moisture supply in 1988 -1991, and a shortage of heat and moisture in 1986. lack of heat with sufficient moisture in 1987. July was moderately warm with a lack of precipitation in 1988 and 1989, warm with frequent rains in 1987, 1990 and 1991, and cool and dry in 1986. August and early September turned out to be warm and dry in 1987 and 1991, cool with frequent precipitation in 1988 and 1989, and warm and humid in 1986.

In 1992 there was a prolonged and cold spring, a cold June, rainy August. The year was generally unfavourable and atypical for the zone. In 1993, the first and second decades of May were characterized by heavy precipitation. The third decade was favourable for carrying out spring fieldwork. The average temperatures and precipitation in the summer months were close to the long-term average.

From August 25 to September 23, precipitation fell almost daily. The growing season of 1994 was characterized as warm and dry. The average monthly air temperature in the summer months was slightly higher than the long-term average. Precipitation fell in May – 91%, June - August - 77% of the norm. The first autumn frost came on September 3. Overall, the year was atypical for the zone. In 1995, spring started very early. The snow cover came off on April 10. May was cold, windy and dry. Half of the average monthly precipitation fell. The first and second five days of July were cold with an average air temperature of 8.7 and 10.7°. During the same period, a large amount of precipitation fell - 20.5 and 32.8 mm, as a result of which no shoots of thermophilic crops were obtained, and the weather was hot and dry in the future

In 1996, the snow cover came off on April 17. The air temperature in May and June was at the level of the long-term average, while precipitation fell by 214% to the average monthly level. July turned out to be hot and dry. In August, the average monthly air temperature was only 12.6 degrees, and 104.3 mm of precipitation fell (with a norm of 66.3 mm). Lack of heat and excess moisture delayed the ripening of many crops. In 1997, the snow cover came off on April 10, which was 10 days earlier than the previous one. The spring was warm, and the average temperatures in April-May were above the long-term ones. Precipitation in May practically did not differ from the average long-term precipitation. In the summer months, the average temperatures were below the long-term average. Precipitation in August amounted to 114.4 mm with a norm of 67 mm, which significantly affected the yield. In general, the growing season turned out to be cool with an uneven distribution of precipitation. Thus, according to the totality of hydrothermal indicators, 1986 can be considered the most favourable for thermophilic plants, the least in 1989 and 1995. The years 1986, 1991, 1993 and 1995 were favorable for cold-resistant plants.

2.2. Research Methods

2.2.1. Obtaining a Stimulator – Activated Water

Magnetization was performed by exposing a 1 cm diameter stream of tap water with a flow rate of 3-4 m/s to a magnetic field of 0.2 T. Tap water electrolysis was performed in a 1.5 l diaphragm electrolyzer. The dielectric body was made in the form of a cube. A flat graphite anode and a titanium cathode; a neutral (canvas) diaphragm were used. Before the experiment, the electrode surfaces were cleaned of salt deposits. During electrolysis, a constant voltage of 200 V was applied to the electrodes for 15 min. Two modifications of EHA were obtained: anolyte (an acidic solution with a positive Eh) and catholyte (an alkaline solution with a negative Eh) by mixing them into a mixture of anolyte and catholyte. Thermal degassing of tap water was performed by heating to 99° in an open vessel, followed by rapid cooling without air access to 20°. The cooling time corresponded to the time it took to heat the water to the appropriate temperature.

2.3. The Effect of Pre-Sowing Treatment With Activated Water of Vegetable Seeds on their Yield in Protected Soil

2.3.1. Cucumber

Pre-sowing treatment of cucumber seeds of the Relay hybrid was carried out in the laboratory of seed science of the Siberian Botanical Garden of Tomsk State University. The seeds were left dry (control 1), treated with tap water (control 2), magnetized, degassed water, as well as catholyte, anolyte and their mixture (1:1) for two hours at a ratio of seed and liquid volumes of 1:5. Germination energy was taken into account after 3 days, germination after 7 days. The morphological analysis of the seedlings was performed on the tenth day after sowing.

Production tests of the effect of pre-sowing treatment of ECAW seeds on cucumber yield were carried out in greenhouses of the Tomich state farm on an area of protected soil of 2.2 hectares in 1986. The seeds previously soaked in activated water were slightly dried and sown in a moderately moist peat substrate with the addition of rotted sawdust, manure and mineral fertilizers * December 17). Control and comparative observations were carried out in fourfold repetition on plots with an area of 8 m, on which 20 plants were planted. The yield was taken into account by separate collection and weighing of fruits from each plot (46 separate collections were carried out during the accounting period). The quality of the fruits was analyzed in an agrochemical laboratory.

2.3.2. Tomato

In 1987, the Tomich State Farm company tested the pre-sowing treatment of ECAW tomato seeds of the Verliotka variety. On June 18, the seeds were treated for an hour with an ECAW at a ratio of seed and liquid volumes of 1:5. On June 19, the seeds dried to flowability and were sown into the prepared substrate. On August 13, the seedlings were planted in a permanent place. The area of the plot was 8 m. The repeatability of the experiment was fourfold. The yield was taken into account by separately collecting and weighing fruits from each plot. The quality of the fruits was analyzed in an agrochemical laboratory. During the elimination of the experiment, the morphological parameters of plants were measured. Similar studies with the Tortilla variety were carried out at the Tomich state farm

Production tests were carried out in the greenhouses of the Kuzovlevsky state farm on an area of 4.4 hectares. Tomato seeds of Claudio and Solveig varieties were treated with ECAW for an hour, and then they were slightly dried and poured into moderately moist soil. Control and comparative observations were carried out fourfold on plots with an area of 8 m, on which 20 plants were planted. Morphological parameters of plants were regularly recorded. The yield was taken into account by separately collecting and weighing fruits from each plot. The analysis of fruit quality was carried out in the agrochemical laboratory of the state farm.

RESEARCH RESULTS

The Effect of Pre-Sowing Treatment of Tomato Seeds With Activated Water on Plant Productivity

We have researched the effectiveness of the pre-sowing stimulation of tomato seeds of the Verlioka variety in production conditions. It was found that stimulated plants from the first days of development differed in larger sizes – the height of seedlings, the area and the number of leaves (Table 2). Probably, these differences would have persisted and intensified at a later date, but the phytotechnics adopted in greenhouses (pinching and pasting) led to their levelling.

Table 2. The effect of pre-sowing treatment of ECAW tomato seeds of Verlioka sort on plant habitus

Variant	Control	Electrochemically activated water
XI*	60,4±5,9	68,9±8,6
X2	7,7±1,0	8,6±0,8
X3	60,5±5,9	69,0±3,0
X4	312,7±10,0	443,1±8,2
X5	76,9±7,7	78,2±1,3
X6	13,5±1,5	15,4±1,3
X7	299,0±22,7	308,8±23,4
X8	12,7±1,3	11,7±1,2

**Note.* XI—plant height, mm; X2—width, mm; X3 — length of cotyledons, mm; X4 is the area of cotyledons at the seedling stage, X5 is the height of the plant, cm; X6 - number of leaves at the beginning of flowering pcs.; X7—plant height, cm; X8 - number of leaves during the liquidation of the experiment, pcs.

A significant increase in the number of brushes, flowers, ovaries and fruits was found (Table 3).

Table 3. The effect of pre-sowing treatment of ECAW tomato seeds of the Verlioka variety on the structure of plant productivity (data obtained during the liquidation of the experiment)

Variant	X9	X10	XII	X12	X13	X14	X15
Control	12,7 ±1,3	9,5 ±2,0	8,4 ±1,8	6,0 ±2,0	1,1±0,1	1,6 ±0,1	23,4 ±0,4
Electrochemically activated water	11,7±1,2	11,4±2,0	5,7 ±2,0	12,8 ±3,4	1,5 ±0,3	4,0 ±0,6	31,3 ±0,5

*Note: X9 – number of brushes, pcs.; XI0—number of flowers, pcs.; X11—number of ovaries, pcs.; XI2—number of fruits, pcs.; XI3 - mass of one plant, kg; X14 - weight of fruits per plant, kg; XI5 - weight of one fruit, g.

An increase in the number of generative organs and the average fruit weight of the tomato variety Verlioka indicates more intense physiological processes in stimulated plants. It should be noted that in case of (emergency) temperature fluctuations in greenhouses, they showed great stability.

Figure 1. The influence of pre-sowing treatment of tomato seeds with activated water on plant productivity

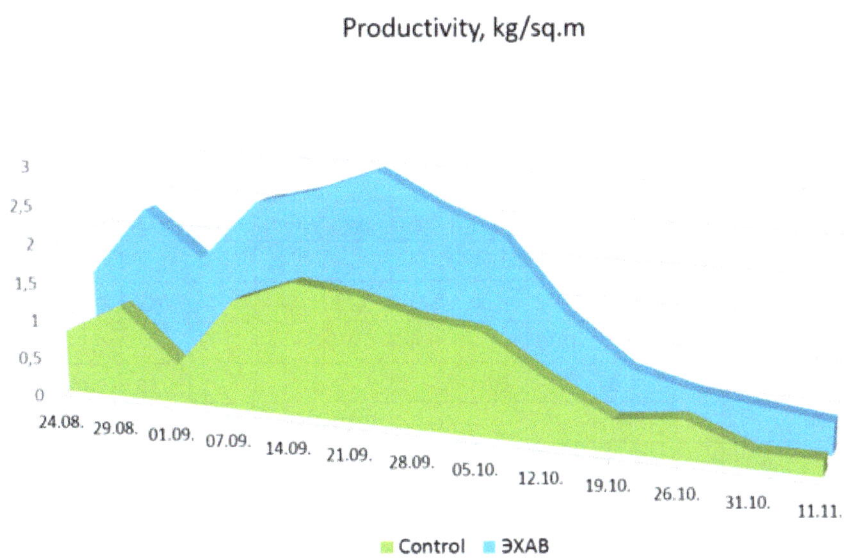

Biochemical analysis showed an improvement in the quality of the fetus under the influence of stimulation. The dry matter yield was 9-13%, the sugar content was 5-9%, vitamin C was 8-9%, and the nitrate content decreased by 7-14% relative to the control. In addition, plants treated with activated water significantly reduced the number of non-standard fruits due to mild late blight. Correlation analysis revealed certain patterns in the development of tomato plants (Table 4). The height of the seedlings positively determined the height of the plant with an average force at the end of the experiment, which allowed it to be used as a diagnostic sign.

The number of fruits was determined by the number of leaves. The height of the plants was significantly related to the number of brushes. The number of flowers was inversely related to the number of leaves during flowering. The number of fruits was directly proportional to the number of leaves and the number of brushes, fruits from one plant, the height of the plant at the beginning of flowering, brushes on the plant and inversely proportional to the number of ovaries. The weight of the fruit depended on the height of the plant at the beginning of flowering, and the number and number of flowers. An average negative relationship between the height of seedlings and the weight of the fruit was also noted. Taller plants formed small fruits.

Table 4. The effect of pre-sowing stimulation of tomato seeds of the Verlioka variety with activated water on plant development (correlation analysis, $p < 0.05$)

	XI	**X2**	**X3**	**X5**	**X6**	**X7**	**X8**	**X9**	**X10**	**X12**	**X13**	**X14**	**X15**
X1				-0,34		0,34				0,16	0,14	0,24	-0,42
X2						-0,36	0,21	-0,40	-0,40		-0,26	0,05	
X5			0,44		-0,36		-0,36				0,03	0,37	-0,24
X6						-0,36				-	0,15	0,16	-0,10
X7	0,34							0,32			-0,02	0,111	0,05
X8		-0,37		-0,35	0,32					-	0,26	0,04	0,20
X9		-0,40											
X10					-0,40								
X11												-0,45	
X12		-0,40					0,57	0,53			0,23 1	0,24	-0,07
X13	-0,39						0,26		0,35	0,23		-0,16	-0,03

continued on following page

Table 4. Continued

	XI	X2	X3	X5	X6	X7	X8	X9	X10	X12	X13	X14	X15
X14		-0,5		0,37			0,04	0,33		0,24	-0,16	—	-0,23
X15	-0,42	-0,20		0,47			0,20	0,40	0,36	-0,07	-0,03	-0,23	

Note. XI — plant height, mm; X2 — cotyledon width, mm; XS — cotyledon length, mm; X5 — plant height, cm; X6 — number of leaves, pcs., — as of August 12 (beginning of flowering); X7 — plant height, cm; X8 — number of leaves, pcs., — when eliminating the experiment on October 16; X9 — the number of brushes, pcs.; X10 — the number of flowers, pcs.; XI1 — the number of ovaries, pcs.; X12 — the number of fruits, pcs.; X13 — the weight of one plant, kg; XI4 — the weight of fruits from one plant, kg; XI5 — the weight of one fruit, g.

The described patterns indicate that the accelerated growth of the habitual sphere contributed to the earlier formation of the crop. This was confirmed by the increase in the yield structure of ripe fruits. The proportion of red fruits during the elimination of the experiment was 1% on control plants, and 51% on stimulated plants (with an overall increase in the number of fruits per plant). The data of the correlation analysis, which revealed a medium-strength relationship between the size of plants at the initial stages of development and the number of fruits, can be considered direct evidence of the legitimacy of predicting plant yields by the strength of their initial growth. Note that previously numerous literature data illustrated this possibility, but the analysis did not go beyond the statement "more powerful seedlings form productive plants". We have also obtained a mathematical proof of this relationship.

In addition, it was found that pre-sowing treatment influenced such indicators as the size of seedlings in early development (growth strength), and the number of brushes in a fruiting plant. The estimated share of the stimulation effect was 24-27% in terms of growth strength and 25% in terms of the number of brushes (Fig. 2).

Figure 2. The proportion of the effect of pre—sowing stimulation of tomato seeds of the Verlioka variety with activated water on the number of generative organs at the end of the experiment: X9 — the number of brushes, pcs.; X10 — the number of flowers; XII — the number of ovaries, pcs.; XI2 - the number of fruits, pgg.

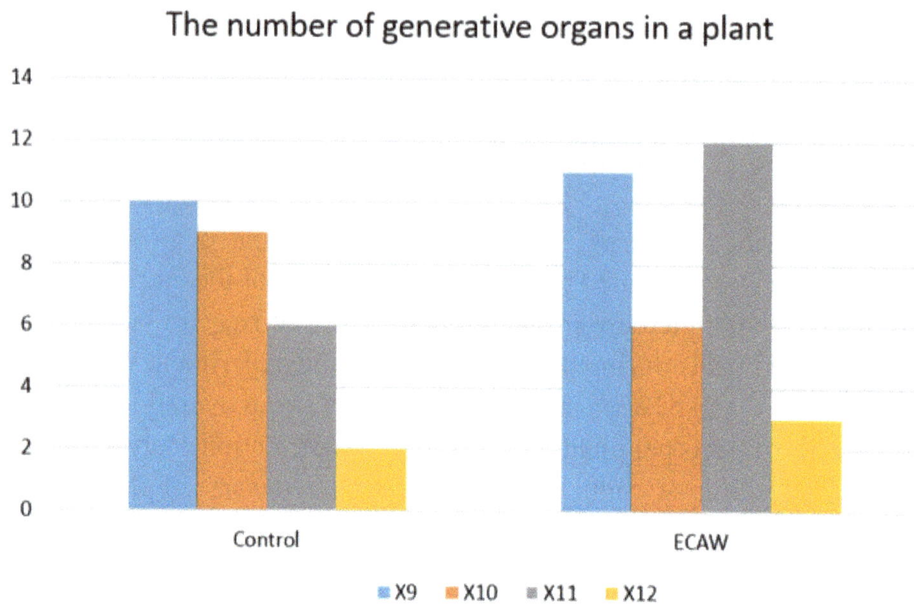

Regression analysis allowed us to establish the high sensitivity of tomato plants to the effect of stimulation. For all the above regression equations, verification using variance analysis showed their adequacy at the achieved significance level of no more than 0.02.

X8 = 2,97 + 0,42-X6, R^2 *=0,619, p = 0,008;*

X10 = 0,64 + 0,5-X15, R^2 =0,47, *p*= 0,05;

X12 = 0,25+ 0,44-XI4, $R^{2\,2}$ =0,835, *p* = 0,02;

X14 = 0,05-0,15-X9 = 0,04 + 0,23-X12, R^2 =0,910, p = 0,05.

Here R^2 is the coefficient of determination.

We have carried out a similar treatment of Revermun tomato seeds with a mixture of catholyte and anolyte at the state farm company Tomich. Starting from the very first phases, the stimulated plants were characterized by a faster rate of development. They formed more powerful leafy ones, on which a greater number of vegetative organs developed. There was a noticeable shift towards brown fruits in the crop structure (Figure 3).

Figure 3. The structure of the crop of tomato plants of the Revermun variety during the elimination of the experiment (average values according to measurements of forty plants carried out during mass fruiting; differences are significant at the significance level of 5%)

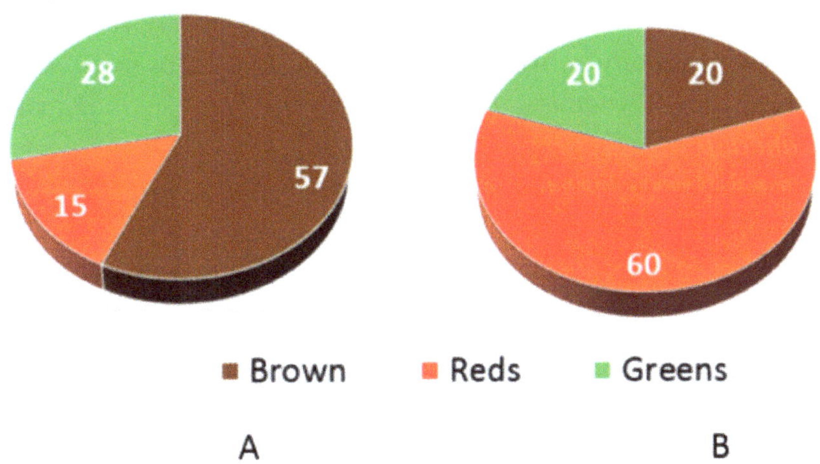

In the greenhouses of the Kuzovlevsky state farm in the Tomsk region, we carried out the introduc-tion of pre-sowing treatment of tomato seeds of the Tortilla variety with a mixture of catholyte and anolyte. The increase in yield was 44% (Table 5, Figure 4). The quality of the products met the appro-priate requirements.

Table 5. The influence of pre-sowing treatment of cucumber seeds on plant productivity, kg/m²

Modification of activated water	February	March	April	May-June	Total yield	
					kg/m²	% to control
Tap water (control)	0,22	3,56	5,03	6,23	15,00	100
Magnetized	0,25	3,96	4,59*	7,14	15,9	106
Degassed	0,47*	4,86*	5,98*	7,63*	18,9*	126
Anolyte	0,38*	4,05	5,76*	7,21*	17,4*	116
Catholyte	0,37*	4,40*	5,64*	7,93*	18,3*	122
A mixture of anolyte and catholyte	0,53*	5,37*	6,13*	7,97*15,9	20,0*	133

Figure 4. Effect of pre-sowing treatment of tomato variety Tortilla seeds with electrochemically activated water on plant yield (average data for greenhouses): a - dynamics; b - final yield indicator. Differences are significant at $p < 0.05$

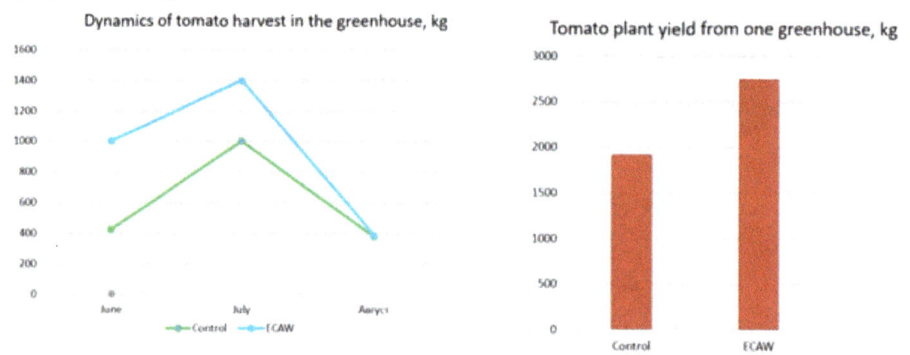

A B

Thus, based on the results of the work done, the following conclusions can be drawn.

1. Pre-sowing treatment of tomato seeds with electrochemically activated water helps to increase the growth force, accelerates and enhances the development of the vegetative sphere, and increases plant productivity and the quality of marketable products.

2. The strength of growth is significantly positively correlated with the yield of plants and can serve as a prognostic sign when evaluating the effectiveness of a particular effect.

3.1. The Effect of Pre-Sowing Stimulation of Cucumber on Plant Productivity

We have conducted a production test at the state farm company Tomich. The effectiveness of pre-sowing stimulation of cucumber seeds on the Relay with various modifications of activated water has been studied, and experiments have shown a steady effect of yield growth in almost each of the 46-50 collections. Phenological observations revealed an earlier (2-4 days) onset of the phenophases of flowering and fruiting in experimental plants, the yield of cucumber in three months in all experimental variants significantly exceeded the control (Table 5). The dynamics of the yield growth of cucumber plants are shown in Figure 5.

Figure 5. Dynamics of productivity growth of cucumber plants in the production experience

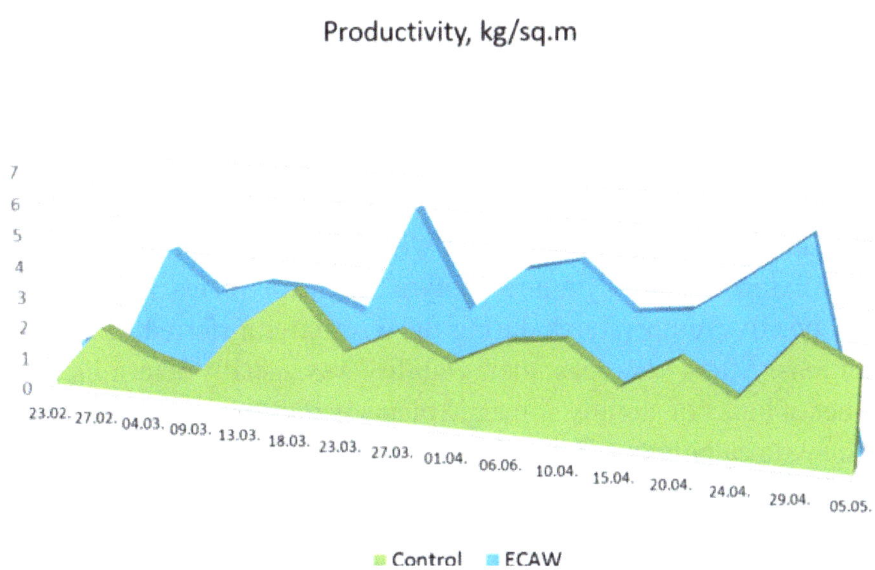

The maximum effect was achieved in the versions with degassed water and a mixture of catholyte and anolyte, the minimum effect was achieved with magnetized water. In some variants, an improvement in fruit quality was noted: an increase in the content of dry matter, vitamin C, and a decrease in the concentration of nitrate nitrogen (Table 6).

Table 6. The effect of pre-sowing treatment of cucumber seeds on the biochemical composition of fruits

Modification of activated water	Dry matter, %	Nitrate nitrogen, mg/kg	Ascorbic acid, mg/kg	Fruit weight, g
Tap water (control)	2,46	259	11,24	221,1
Magnetized	2,91	254	12,37*	232,1*
Degassed	3,40*	226*	12,25*	235,7*
Anolyte	3,47*	242*	12,43*	238,4
Catholyte	3,67*	241*	12,36*	229,1
A mixture of anolyte and catholyte	3,80*	171*	12,65*	237,4*

*Differences at a significance level of 5%

Thus, of the studied modifications of activated water, degassed water and a mixture of catholyte and anolyte have the best stimulating properties for cucumber plants, while magnetized water has the worst.

We also conducted an additional study of the effect of watering cucumber seedlings ECAW on the frost resistance of plants. To do this, the plants were watered daily for three weeks with tap water (control), catholyte and a mixture of catholyte and anolyte. Then the plants were exposed to low positive temperatures during the day and transferred to favorable conditions. After two weeks, their viability was visually determined. The number of intact or slightly damaged plants in the control was 36%, in the variant with catholyte — 50%, with a mixture of catholyte and anolyte — 57% Consequently, watering plants with activated water increased their resistance to low positive temperatures.

CONCLUSION

Summing up the results of our research and the literature data of other authors, the mechanism of action of activated water can be imagined as follows. Physical factors (magnetic field, temperature, vacuuming. Ultraviolet irradiation, ultrasound treatment, etc.) contribute to a change in the physico-chemical properties of the water-gas-salt system. When interacting with seeds, they trigger a variety of metabolic processes and contribute to the activation of seed germination, increase germination energy and, in some cases, their germination. The most important manifestation of the stimulating effect of any factor of a physical and chemical nature is the strengthening of organ-forming activity at the early stages of plant development.

Strengthening plant nutrition by stimulating rhizogenesis provides a more effective mineral nutrition for plants and manifests itself in an increase in the number of vegetative and generative organs. Activated water is not only an effective but also a cheap stimulator of plant growth and development. This is especially important when introducing technology into the manufacturing sector. In addition to its high efficiency, it preserves the ecological purity of the soil and agricultural products, since in the process of relaxation, the properties of water return to the initial level, and no additional substances are introduced into the environment.

In our large-scale production experiments, the high efficiency of using electrochemically activated water has been shown to increase the yield of tomatoes and cucumbers. In all experiments, an increase in the overall yield was noted, and, most importantly, the early, most profitable yield of vegetable products. An important point is to improve the quality of agricultural products under the influence of pre-sowing (or pre-planting) stimulation.

Undoubtedly, experiments on the effect of activated water are promising not only for vegetable growing, but also for horticulture, floriculture, and medicinal plant growing. The main thing is to control the physicochemical parameters of water during its treatment by physical factors. This will ensure a scientific approach, reproducibility of experiments and repeatability of results, as well as avoid profanation and discrediting of technology, which was at the beginning of the study of biological and medical properties of water due to its surface use by amateurs.

Prospects For The Development Of Research are associated with the expansion of the objects of influence (species, varieties of cultivated plants) of activated water and other growth regulators, clarification of the mechanisms of influence with the aim of their optimization, as well as the expansion of the range of goals and objectives solved with their help.

REFERENCES

Afzal, I., & Javed, T., Amirkhani. (2020). Modern technologies of the sowing system: Application of coatings on seeds to increase seed yield and yield. *Agriculture, 10*, 526. DOI: 10.3390/agriculture10110526

Benazzuk, S., Jazuli, Z.E., Lutts, S. (2019). Vermicompost filtrate as a promising means for priming and rejuvenation of germinating Brassica napus seeds treated with salt. *Communes. Soil science. Vegetable anal.*, 50, 1344–1357.

Hinojosa-Davalos, J., Arias-Rios, EV, & Varela-Hernandez, J.J., Cardona-Lopez, Massachusetts, Orozco-Muniz, R., Cabrera-Diaz, E. (. (2020). Thermal and chemical treatment of alfalfa seeds (Medicago sativa) and broccoli (Brassica oleracea var. italica) before and during the germination process: An approach to salmonella germination. *Journal of Food Protection, 83*, 82–88. DOI: 10.4315/0362-028X.JFP-19-272 PMID: 31851547

Javed, T., Ali, M.M., Shabbir, R., Gall, S., Ali, A., Khalid, E., Abbas, A.N., Tarik, M. (2020). The emergence of rice seedlings under the influence of various factors and seedling with potassium nitrate. J. *Adj. Res. The science of plants*, 1, 65-75.

Kadena, M.B., Preston, G.M., Van der Hoorn, R.A., Flanagan, N.C., Townley, O.N., Thompson, I.P. (2018). Increasing the activity of cinnamon essential oil by encapsulating nanoparticles to combat seed pathogens. *Indian crop harvest*. 124, 755-764.

Komarov, A.A. (2004). The nature of humic preparations and their effect on cucumber plants, 7, 22 - 23.

Komarov, A.A. (2005). Physiologically active substances reduce the accumulation of nitrates in Chinese cabbage *Potatoes and Vegetables*, 3, 24 - 25.

Mustafa, , GShehzad, , M. ATahir, , M. H. NNawaz, , FAkhtar, , GBashir, , M. AGhaffar, , A. (2022). Pretreatment with chitosan arbitrates physiological processes and antioxidant defense system to increase drought tolerance in alfalfa (Medicago sativa L.). *Journal of Soil Science and Plant Nutrition, 22*(2), 2169-2186.

Pandiselvam, R., Mayookha, V. P., Kothakota, A., Sharmila, L., Ramesh, S. V., Bharathi, C. P., Gomathy, K., & Srikanth, V. (2020). Ozone treatment method for seed germination - a systematic review. *Ozone Science and Engineering*, *42*(4), 331–346. DOI: 10.1080/01919512.2019.1673697

Pasko O.A. (1991). Acceleration of seed germination when treated with electrochemically activated water. *Physiology and biochemistry of cultivated plants*, 23 (6), 552 – 556.

Pasko, O.A. (2010a). The growth and development of plants stimulated by electrochemically activated water. *Bulletin of the B.S.S.A. named after V.R. Filippov scientific and theoretical journal*, 3,54-59.

Pasko, O. A. (2010b). Stimulation of vegetative reproduction of ornamental plants using electrochemically activated water. *Phytodesign in modern conditions materials of the International Scientific and Practical Conference*, Belgorod: BelSU Publishing House, 146-147.

Pasko, O. A. (2010c). Physico-chemical changes in tap water during its treatment in various ways. Property changes of piped water treated with different methods. *Water: Chemistry and Ecology*, *7*, 40–45.

Pasko, O. A. (2011). The influence of pre-sowing stimulation of cucumber seeds on yield. *Agricultural Sciences*, *8*, 20–22.

Pasko, O. A. (2011b). Stimulation of vegetative reproduction of plants with sodium humate Innovative aspects of peat extraction, processing and application. *Proceedings of the International Conference dedicated to the 115th anniversary of the National Research* Tomsk Polytechnic University, Tomsk: TPU Publishing House, 122-125.

Pasko, O. A. (2011c). The effect of presowing stimulation of cucumber seeds on yield. Influence of presowing stimulation of cucumber seeds on productivity. Agricultural science. Scientific. *Theoretical and Industrial Journal*, *8*, 20–22.

Pasko, O. A., & Gomboev, D. D. (2011). *Activated water and the possibilities of its application in crop production and animal husbandry*. Tomsk Polytechnic University Publishing House.

Pasko, O. A., & Ketskalo, V. M. (1991). The influence of electrochemically activated water on seed germination. *Plant Physiology*, *38*(2), 381–385.

Rybinsky, V. (2001). The effect of laser beams in combination with chemomutagen (MNU) on the variability of traits and mutation frequency of spring barley. *International Agrophysics*, *15*, 115–119.

Sakala, E., & Demchuk, A., Grzhi, Oe E., Proeba-Bialchik, U., Scheisner, H. (2012). The effect of pre-sowing laser irradiation of seeds on the properties of sugar beet. *International Agrophysics*, *26*, 295–300. DOI: 10.2478/v10247-012-0042-6

Samarah, , N. HAl-Quraan, , N. AMassad, , R. SWelbaum, , G. E. (2020). Treatment of bell pepper (Capsicum annuum L.) seeds with chitosan increases chitinase and glucanase activities and enhances emergence in a standard cold test. *Scientia Horticulturae, 269*, 109393.

Shera, B., Scholz, V., Ireshova, Yu., Hong, J., Yulak, J., & Sheri, M. (2021). Investigation of non—Thermal plasma treatment on seed germination and early growth of legumes - review. *Plants*, *10*, 1616. DOI: 10.3390/plants10081616 PMID: 34451662

Sivasubramaniam, K., Gita, R., Sujata, K., Raja, K., Sripunitha, A., & Selvarani, R. (2011). Seed sowing: Triumphs and aDWersities. *Madras is. Agricultural Journal*, *98*, 197–209.

Stadler, L. J. (1928). Mutations of barley induced by X-rays and radium. *Science*, *68*(1756), 186–187. DOI: 10.1126/science.68.1756.186 PMID: 17774921

Talanova, L. A. (2006). Justification of the effectiveness of methods for treating cucumber seeds and plants with magnetized water and humic acids. Abstract of the dissertation for the degree of candidate of agricultural sciences in the specialty 06.01.06, Vegetable growing. Ryazan, 27.

Tretyakov, H.N. (2005). Physiology and biochemistry of agricultural plants. *Adaptation and stability of plants.* Moscow, Kolos, 640.

Vasiliev, V.V., Slesarenko V.V., Gurskaya, T.A., Primachev V.I. (2010). Activated water increases fruiting of vegetable crops. *Potatoes and vegetables*, 6, 27.

Zelepukhin, I.D., Pasko O.A., Aksenova E.A. (1988). *Application of activated water in agriculture and biotechnology*. Tomsk: Science and production.

Chapter 7
Sustainable Agriculture Processes Through Additive Technology: Circular Economy

T. Kumaresan

Department of Mechanical Engineering, PSG Polytechnic College, Coimbatore, India

S Prabhu

Department of Mechatronics Engineering, PSG Polytechnic College, Coimbatore, India

Avni Suresh Thakkar

School of Management, Shri Ramdeobaba University of Engineering and Management, Nagpur, India

Sampada Viraj Dravid

Department of Mechanical Engineering, Vishwakarma Institute of Technology, Pune, India

V. Krishnamoorthy

https://orcid.org/0009-0005-9250-9908

Department of Management Studies, Kongu Engineering College, Perundurai, India

DOI: 10.4018/979-8-3693-4392-0.ch007

ABSTRACT

Additional technologies commonly referred to as 3D and 4D printing, is emerging as a transformative technology in sustainable agriculture in the structure of the circular economy. Local of agricultural equipment, parts of equipment, and local, enabling production on demand, reduces AM waste, reduces carbon emissions and increases resource efficiency. This chapter explores how innovative AM technologies support clause-loop agricultural systems by reducing fake, customization and supply chain dependence. It also discusses the integration of renewable and bio-based materials in AM processes, promoting regenerative methods. Eventually, AM provides a scalable way to elastic, eco-efficient and circular agricultural value chains for the future.

INTRODUCTION

Agriculture, the backbone of human culture, the rising global challenges, such as the backbone, weather change, resource depletion and insecurity of food, are taking major changes. Traditional agricultural practices, when productive in the past, are increasingly criticized for their environmental influences, inaccuracies and linear resource consumption models. As societies turn to more durable and circular approaches, the integration of advanced technologies in agriculture has given significant traction. In these technologies, additive manufacturing (AM), commonly referred to as 3D printing, emerged as a promising capable of innovation in durable agricultural systems. Customized, efficient, and locally produced the demand for the demand for the equipment and components, closely adjusts with the principles of the AM circular economy-reducing the waste, the use of the resource and enabling regeneration to the Agro-value chain (Lu et al., 2024).

Additional manufacturing is basically different from traditional subtractive production techniques. Instead of cutting materials from large blocks, AM creates a layer-by-layer using digital designs and range of digital designs and feed stocks, including digital design and plastic, metals and more and more, bio-based or biodegradable materials. This method provides many internal benefits: it reduces material waste, allows for rapid prototyping and repetition, supports decentralized production, and enables product design to suit specific agricultural needs. For smallholder farmers and agricultural innovations, especially in remote or resource-limited regions, AM presents the opportunity to create the necessary equipment, spare parts, irrigation components, seed planters and clay testing equipment, without depending on long supply chains or expensive imports. This capacity significantly improves the elasticity and self -reliance of local farmer communities (Balla et al., 2022).

In the context of the circular economy, agriculture must be concluded and transition to systems that revive with linear input-output models, installation and resource-rupee. The circular economy promotes the use of materials and products for as long as possible, the maximum value from them, and revives natural systems. Additional manufacturing farms meet these goals by enabling modularity, repairability and upgrading in equipment, while also promoting the use of biodegradable or recycled materials in fake processes. For example, 3D-printed tools made from recycled plastic or agricultural residues reduce not only environmental influences, but also create new economic opportunities in rural circular ecosystems (Xu et al., 2024).

Moreover, it increases the precision and efficiency of agricultural operations. With digitally operated design and production, Farm Tools can be optimized for specific crops, ground conditions or climate factors. This level of customization supports precision agricultural practices that reduce inputs such as water, fertilizer and pesticides while maximizing productivity and durability. In addition, the digital nature of AM allows to share open sources of agricultural designs and innovations. The platform where farmers and designers can solve collaborative problems and accelerate the diffusion of sustainable technologies in regions (Malik et al., 2022).

Other digital and biological innovations - such as Internet Th F -Things (IoT), Artificial Intelligence (AI) and artificial life. No converting to additive manufacturing expands its transformative potential in agriculture. For example, AM can be used for sensor housing or drone components to adapt to certain monitoring tasks or for prototype cultivated farming modules for controlled-in-environment agriculture. In Agriculture, AM is being invented to create bifactors or lab-on-a-chip devices for rapid genetic or clay analysis. These applications highlight how AM serves as a crucial infrastructure technology in widespread smart and circular agricultural patterns (Parmar et al., 2023).

However, despite his promise, A.M. in agriculture. Adoption is not without challenges. Technical limitations such as physical strength, printer scalability and durability in the state of rigid field need attention. Moreover, there may be significant obstacles to digital literacy, design skills, and reliable electricity or internet connections. The environmental impact of some AM materials, especially petroleum -based plastic, must be carefully managed to ensure configuration with stability goals. To fully realize the possibility of AM in circular agriculture, stakeholders - including governments, research institutes, agricultural expansion services and private sector artists - invest in capacity building, open innovation ecosystems and support policy structure (Khan et al., 2022).

This chapter discovers a multilateral role of additive manufacturing in enabling circular economy transition to agriculture. It is detecting current innovations, materials and cases of use, showing how AM contributes to stability, efficiency and elasticity in the food production system. Through case studies and analysis, it examines the socio-economic and environmental effects of integrating AM technologies into farming methods. The debate also addresses future directions, including biodegradable feed stocks, the role of distributed manufacturing networks and digital design communities and shaping the resurrected agricultural future (Soni et al., 2022).

In conclusion, such as the global agriculture sector, ecological stewards and social equality try to balance productivity, offering a crucial lever for technologies such as additive manufacturing. By supporting local, flexible and resource-efficient production, AM empowered to be an active participant in a more circular and durable food system. The integration of AM in agriculture is not just a technical shift, but the innovation of systems-level is how we produce, consume and revive in harmony with the planet. It is necessary to understand its potential and difficulties to take advantage of its benefits in the search for food security, environmental elasticity and rural development (Kabato et al., 2025).

Objectives

The purpose of this chapter aims to explore the transformative role of additive manufacturing (AM) to further sustainable agriculture in the structure of the circular economy. He wants to identify how AM technologies can reduce content waste, increase resource efficiency and promote local, customized agricultural production. Objectives include examining the innovative applications of AM in farm operations, evaluating the use of bio-based and recyclable materials, and evaluating its contribution to elastic and regenerative agricultural systems. In addition, this chapter aims to highlight the opportunities for AM to integrate with digital agriculture and identify the challenges and policy considerations needed to support its wider adoption in rural and peasant communities.

Figure 1. Foundations of additive technology in agriculture

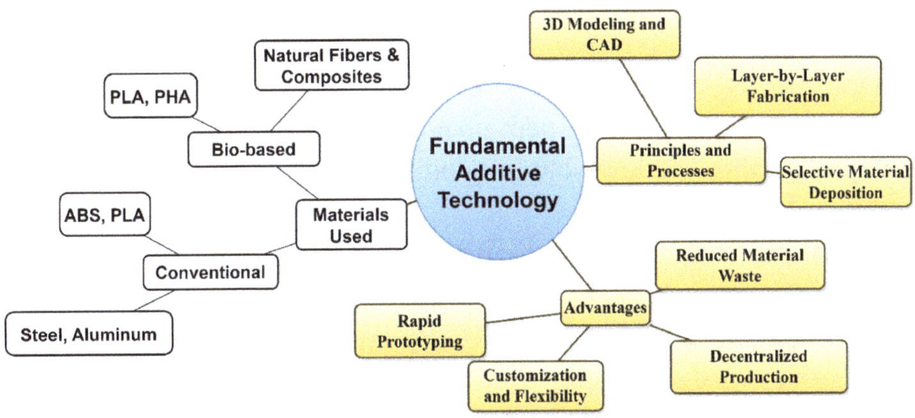

FOUNDATIONS OF ADDITIVE MANUFACTURING

Principles and Processes

Additional manufacturing (AM), often synonyms of 3D printing, represents the example of how the products and components are designed, developed and manufactured. At its origin, AM refers to a group of technologies that generally make Objects Layer-Layer directly from digital models, using a computer-added design (CAD) software. Unlike traditional subtractive methods, which removes the material from the solid block through machining, cutting or drilling, the material is added to it. This layer-by-layer approach allows complex geometries, lighter weight formats and functional integration that will be expensive or impossible to produce with traditional production methods. Figure provides a comprehensive principle of additive technology in agriculture (Fu et al., 2022).

There are many AM processes characterized by ISO/ASTM 52900 standard. The most common techniques include fused deposition modeling (FDM), which out of thermoplastic filament; Is; Stereolithography (SLA), which uses photopolymerization to cure resin with UV lights; And selective laser sintering (SLS), which fuses the powder material using a laser. These techniques are different in the kind of material, resolution, motion and measure they use. AM processes begin with a digital 3D file that is cut into cross-sectional layers, each of which is printed sequentially until the final Object Budget is formed. As such, AM provides freedom of non-matching design and opens up new possibilities for customization, rapid prototyping and low-volume production.

Materials Used: Conventional vs. Bio-based

There is a significant expansion in the range of materials used in AM, including thermoplastics, photopolymers, metals, ceramics and composite. Traditionally, acrylonitrile butadiene has been chosen due to the ease of printing, structural properties and availability of ABS, PLA and nylon. The use of metal am stain stainless steel, aluminum or titanium powder is usually applicable to high-value areas such as aerospace and medical devices. However, the environmental stability of this traditional material has been called in question, especially related to non-non-ponded waste and residual-consulted feed stocks (Fu et al., 2022).

To overcome these concerns, the development of bio-based and biodegradable material has emerged as an important boundary in additive manufacturing. Bio-based filaments obtained from biopolymers such as agricultural fossils, algae or PLAs and polyhydroxyalkanoates (PHAs) are used for more and more durable programs. For example, the PLA is compostable in Industrial dysfunction and can be obtained from renewable crops such as corn or sugarcane, which makes it a viable eco-friendly option of petrochemical plastic. Other innovations include natural fibers (eg, hemp, wood, rice laughter) for improved mechanical and environmental influence(Sohail et al., 2022).

This bio-based material not only reduces environmental influence but also promotes circular of AM processes. Supporting a resurgent loop in the rural economy, agricultural sub-products can be valued in feed stocks for 3D printing. Nevertheless, bio-based materials still face challenges in terms of consistency, mechanical properties, and scalability, which requires further research and development to complete industrial dysfunction standards.

Advantages over Traditional Manufacturing

Additional manufacturing offers many benefits in terms of especially durable and circular agriculture on traditional production approaches. The most notable benefit is physical efficiency. A.M. It is only where the material is needed, it reduces the waste compared to subtractive manufacturing, which often results in a large amount of scrap material. This is especially important when using expensive or rare materials. Another main advantage is the design flexibility. Makes the design of a very complex, customized design without the need for special tooling or mold. This allows rapid repetitions and prototyping, which is ideal for the development of reference-specific farming equipment or precision agricultural equipment. For example, tools can be designed to match certain crop types, terrain status or ergonomic needs of farmers, especially in smallholders and local settings (Fu et al., 2022; Kabato et al., 2025).

Decentralization of the product is another attractive feature. A.M. Along with the issue of product use, the transport costs, supply chain dependence and reduce carbon emissions. This is especially valuable in remote agricultural regions where the accesses of spare parts or specialized devices are limited. For example, a broken component of a water pump or cider, can be scanned, digitally shared and resigned locally - downtime and operational disturbances are reduced. Moreover, AM only supports the production of time, which adjusts well with a circular economy strategy that emphasizes lean, response and resource-conscious systems. It also facilitates the expansion of the product life cycle through repair and part replacement, increasing the life expectancy of the devices and reducing the need for complete replacement. This not only reduces costs, but also controls the resource extraction and producing waste.

From the point of view of durability, the environmental footprint of the AM may be significantly lower when the renewable energy-powered and combined with recyclable or bio-based inputs. The ability to reuse failed prints, the ability to reproduce waste content in new filament or incorporate biodegradable materials, makes a strong companion in efforts to green the manufacturing value chain. In conclusion, additive manufacturing provides a transformative set of tools and techniques to re -imagine agricultural production and devices through stability and circular lenses. Its ability to produce complex, customized and resource-efficient solutions on demand promises very much for the future of agriculture-to promote self-reliance, waste reduction and environmental stewardship. As AM technology continues to develop, its integration in farming systems will be more and more crucial in eliminating the dual challenges of productivity and planetary health.

Figure 2. Additive manufacturing in agricultural applications

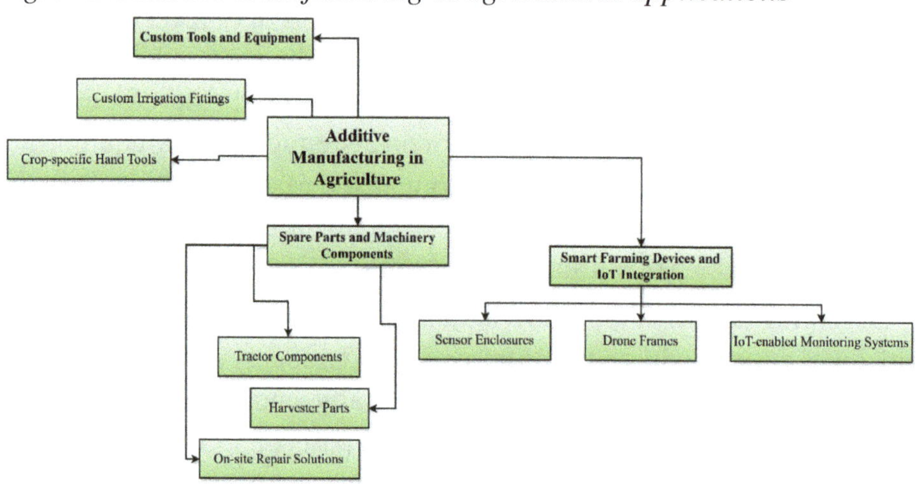

ADDITIVE MANUFACTURING IN AGRICULTURAL APPLICATIONS

Custom Tools and Equipment

The dynamic and diverse nature of agricultural methods often requires reference-specific equipment that is not easily available by the traditional supply chain. Additional manufacturing (AM) offers crisp, cost-effective remedies by enabling the rapid formation and production of custom agricultural equipment corresponding to specific crops, soil or environmental conditions. This customization is especially beneficial for small holder farmers and precision agricultural methods, where one-to-date-fit-all equipment is inadequate. Using digital modeling and open-sun design libraries, farmers and aggrotech startups can quickly develop equipment such as seed planters, pruning equipment, irrigation nozzles and clay samples. These tools can be 3D-printed using combined bio-filaments associated with biodegradable materials such as polylactic acid (PLA) or agricultural waste (eg, hemp fiber, rice rice). Importantly, these tools can be convenient and re -produced using a desk on the spot or on the regional creation hub. 3D printer, both lead time and cost. Figure 2 depicts the use of additive manufacturing in agricultural applications(Khan et al., 2022; Soni et al., 2022).

Customization also plays a crucial role in enhancing ergonomics and user efficiency. For example, handheld tools can be designed with improved grips for different age groups or physical abilities, reducing fatigue and injury. In remote or resource-rear areas, 3D-printed components can be prepared to interface with locally available devices, avoiding issues of consistency. A.M. The recurring nature of the user allows for continuous improvement based on response, facilitates participating design and community -based innovation in agriculture. Moreover, custom tool development arranges well with climate-smart agricultural methods, where local, adaptive equipment is essential to respond to weather patterns, pest pressure and transformation of land health. Tools can be purified to increase microcells or unique crop rotation, efficiency and durability.

Spare Parts and Machinery Components

Agricultural operations often suffer from the availability of spare parts or cost of cost, especially in geographically different areas. Downtime due to broken or boring components can lead to significant yield loss, especially during the critical planting or harvest period. Additional manufacturing tractors address this challenge by enabling demand production of spare parts for tractors, harvesters, irrigation systems and other agricultural machinery. Through 3D scanning and contrasting engineering, obsolete or imported machine parts can be digitally rebuilt and printed locally. Components such as gear wheels, clamps, housing units, pump impellers or nozzles can be produced with a high degree of accuracy using materials such as more reinforced polymers, nylon composite or even metal powder. For less mechanically stressed parts, biodegradable filaments provide an environmental -friendly repair option (Khan et al., 2022; Soni et al., 2022).

This distributed product model reduces dependence on central production facilities and eliminates delays associated with logistics or international supply chain. It also offers economic benefits - production costs, reduction in inventory requirements and minimal waste. In developing regions, initiatives are underway to install rural fabric labs (or "fab labs") that empower communities to produce their own parts and equipment, increasing self -reliance and elasticity. In addition to the replacements, the performance of the components of AM machinery allows optimization. Parts can be redesigned for weight loss, improved aerodynamics or long durability. For example, mesh designs can be integrated to reduce the use of materials without compromising. Innovations like these expand the useful life of machinery and reduce the environmental measures associated with production and disposal.

Smart Farming Devices and IoT Integration

The integration of smart technologies in agriculture - is commonly known as smart farming or precision agriculture - revolutions how the farm is monitored and operated. Additive manufacturing, IoT-enabled devices, plays a major role by providing rapid prototyping and deployment of sensors and housing, which collects real-time data on land status, climate, pest presence and crop health. AM allows the creation of customized sensor enclosures, drone components, GPS mounts and robotic actuators that correspond to specific functions or field environments. Lightweight, waterproof and UV-resistant materials can be selected to increase the durability of the device in agricultural conditions. By combining 3D-printed casings with microcontrollers such as Arduino or raspberry, effective smart systems can be assembled for automatic irrigation, pest check and crop monitoring.

For example, moisture sensor housing can be printed to accurately fit local soil conditions and buried with minimal disruption in the root zone. Weather stations can be equipped with customized wind vans and rain gauge, reducing dependence on imported parts. A drone equipped with 3D-printed structural ingredients can be used for aerial imaging, enabling farmers to detect crop stress and manage more precision inputs. Importantly, AM enhances inter -efficiency and modularity - two main principles in smart farming. Devices can be designed to easily disassemble, repair or upgrade, supporting circular design principles. Moreover, the capacity of rapid prototype repetitions accelerates production development and testing cycles, allowing researchers and innovations to improve efficiency based on field tests (Malik et al., 2022; Parmar et al., 2023).

Democracy of innovation by AM and IOT conversion is especially clear in open-source hardware communities, where farmers and technologies share printable designs and sensor configurations. This promotes peer-to-peer knowledge exchange and reduces obstacles to digital agricultural transformation in the underworld regions. Additional manufacturing is proving to be a transformative power in agriculture, offering varied applications from custom tools and spare parts to smart devices that unite with digital farming systems. Its unique ability to produce customized, demanding and resource-efficient solutions perfectly adjusts with the requirements of the circular and durable agricultural economy. By decentralizing the product and enabling local innovation, AM empowered to be an active participant in the creation of farmers and rural communities in their own equipment and technologies. Access to content, digital skills and printer infrastructure improved, additive manufacturing is ready to be the essential pillar of modern agricultural renewal.

ALIGNMENT WITH CIRCULAR ECONOMY PRINCIPLES

Additional Manufacturing (AM) is more and more known as the main capable of circular economy (CE) in agriculture. CE framework emphasizes resource efficiency, reduction of waste, reuse of materials and thinking of systems - objectives that are equipped to address AM technologies unique. By its ability for precision fake, customization and digital integration, AM supports direct circular methods in the agricultural value chain. This section detects three major configuration points: waste reduction and re -use of materials, localized and demanded on demand, and the thinking of the life cycle by durable design (Hettiarachchi et al., 2022). Figure 3 illustrates the alignment of Additional Manufacturing (AM) and Agriculture with Circular Economy Principles.

Figure 3. Additional Manufacturing (AM) and agriculture alignment with circular economy principles

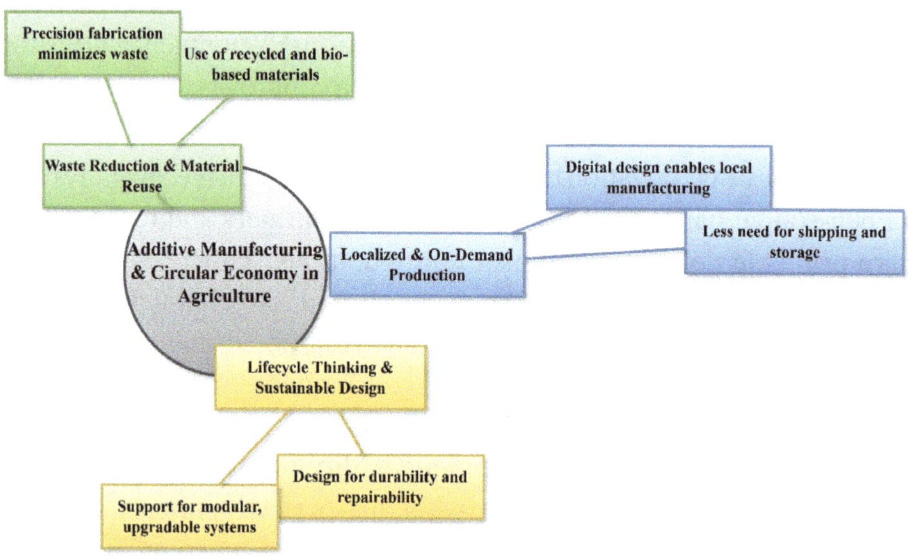

Waste Reduction and Material Reuse

The most important advantage of additive manufacturing in terms of circular is its ability to sharply reduce product waste. Unlike subtractive manufacturing, which removes materials from a solid block - often produces significant scrap - creating a layer of components by layer only using the required material. This near-zero-waste approach is especially beneficial in resource-limited agricultural settings. In addition, AM provides opportunities for repetition and reuse of content (Ciccullo et al., 2021). Failed prints, end of life's end components and agricultural biotests (such as crop residues, chitin or lignin from wood) can be processed in recycled or bio-composed filaments. For example, the initiative is emerging that converts agricultural plastic waste into new 3D printing materials, enabling a closed-loop system. Farmers or cooperatives can grind plastic items or failed prints used in filament, and re-use to print tools or parts, reducing the inputs and reducing the dependence on virgin content.

Localized and On-Demand Production

The main obstacle to durability in traditional agriculture is the central, globalization nature of agricultural machinery and input supply chain, resulting in high transport emissions, no longer lead time and equipment. Additional Manufacturing Uses this, enabling the local product of equipment, parts and equipment - at the stage of use. Local AM setups, such as rural fab labs or community productive spaces, can produce spare parts, smart farming enclosures or customized tools using open-source design. This not only reduces carbon footprints associated with transportation, but also strengthens regional elasticity and self -reliance. Farmers need their needs, when they need - to remove waste from excessive production, warehousing and obsolete inventory. Production on demand also adjusts to supply with real-time demand, thus reducing excessive use of materials and avoiding speculative production. In the climate-sensitive farming environment, this adaptability is crucial-the tool or monitoring devices can be modified for seasonal changes, weather events or pest erosion. Moreover, digital fabrication supports dexterity and rapid prototyping. Designers and farmers can remotely collaborate to repeat new products that react to local ecological or agricultural requirements (Bimpizas-Pinis et al., 2022).

Lifecycle Thinking and Design for Sustainability

Circular economy principles advocate thinking throughout the life cycle of production-from the source of material to the end of the life. Additional manufacturing naturally lend itself for a lifelong design, not just functional but also by designing unmatched products that are durable by design. A.M. Using, designers can embed the dimensions of stability in geometry, structure and materials of agricultural equipment or equipment. Lightweight mesh compositions with AM, topology optimization and multifunctional components are possible, reducing material consumption while maintaining performance. For example, Soil sensor housing can be printed with integrated cable guides and waterproofing features - reducing the need for additional assembly materials or adhesives (Giakoumatos & Tsoulfas, 2025).

Designers can plan to recovery of life by selecting biodegradable materials, enabling part isolation, or by embedding digital QR codes, by giving users information about recycling or reused ways. Considering the full life cycle on the design stage, AM organizes "cradle-to-daedal" thinking-the foundation part of the parishioner economy. Another important contribution of AM is to enable circular data flow. Digital blueprints can be updated, shared and localized for new references without the need to reinstall physical production systems. This obsolete design reduces waste and supports global commons of durable agricultural innovation. Open source repository of farming equipment equipped under Creative C One License is becoming increasingly important for the democratization of sustainable techniques. Moreover, additive manufacturing is well integrated with the life cycle assessment (LCA) tools, which helps evaluate the materials, the use of energy zones and the environmental influence of waste. These tools can guide design preferences to reduce tangible carbon or maximize recyclability - enhancing configuration with circular principles.

Additional manufacturing plays a transformative role in organizing agricultural production and innovation with the principles of circular economy. From reducing material waste and reusing resources, enabling local, demand fabrication and embedding the life cycle thinking in the design, revitalizing the use of AM linear resource, moving for circular methods. AM is set to become a base technology to increase the accession of durable materials, digital design tools and the accommodation of decentralized printing infrastructure, more durable for agriculture, incorporated and circular future.

Figure 4. Innovations and case studies - focusing on how Additive Manufacturing (AM) is transforming agriculture

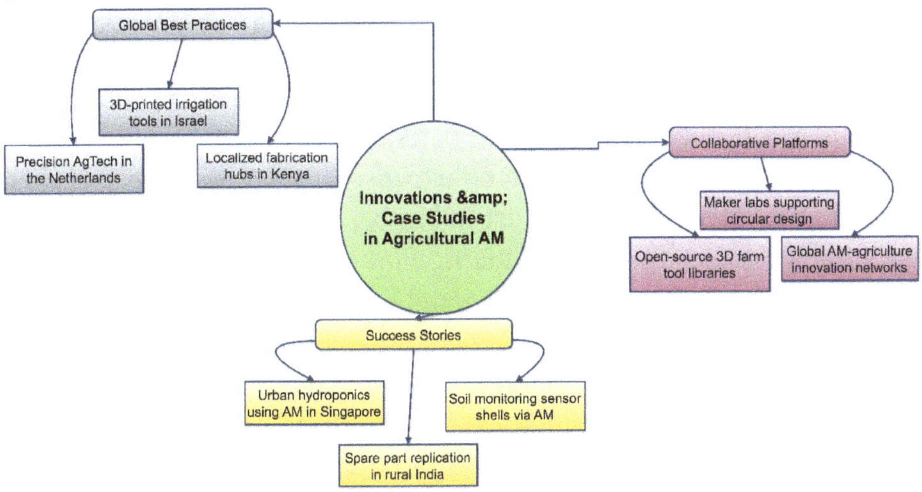

INNOVATIONS AND CASE STUDIES

This department highlights the leading innovations and illustrated case teaching that reflects the changing influence of additive manufacturing (AM) in agriculture. We look for the role of global best practices, success stories of various farming references and the role of collaborative platform to accelerate innovation and circular. Figure 4 showcases innovations and case studies highlighting the transformative impact of Additive Manufacturing (AM) on agriculture (Chowdhury, 2023; Hettiarachchi et al., 2022).

Global Best Practices and Pilot Projects

Over the world, various institutions, research institutes and startups are taking advantage of additive manufacturing to revolutionize agricultural processes. A significant example is the Fabiform project in the Netherlands, which integrate 3D printing with hydroponics and robotics to develop automatic, durable urban farms. A.M. The use of rapid formation of custom parts for optimize for urban environment for sensors, plant holders and irrigation components. Their pilot showed a significant reduction in physical waste and water use, in which showed how precision can support farming at blocked spaces. Similarly, in India, the Tata Trust's initiative has introduced a decentralized 3D printing hub in rural areas. These centers empower farmers to print spare parts for their tractor and irrigation equipment, reducing dependence on expensive and slow supply chains. This initiative also promotes the use of bio-based filaments obtained from agricultural remains, promoting the flow of circular material in peasant communities. In the United States, modular 3D-printed components have been developed for environmental agriculture-controlled environment from the OpenAI initiative at MIT Media Lab. Their innovation is easily focused on replicating, low -cost farming equipment that can be customized for various crops and climates. An example of this project is how A.M. The powered modularity increases scalability and durability in different farm models (Giakoumatos & Tsoulfas, 2025; Tege et al., 2025).

Success Stories from Rural and Urban Farming Contexts

In the context of both rural and urban agriculture, A.M. In rural Africa, for example, Kenya farmers have taken advantage of the 3D-printed parts locally for repairing water pumps and planting equipment. Using recycled plastic and -f-the-shelf 3D printers, communities have drastically reduced costs and machine downtime by increasing agricultural productivity. The innovation of the bottom enhances elasticity and economic freedom in resource-limited settings. In cities like Chicago and Singapore, urban farms have used AMs to develop customized sensor housing and iCal bill farming modules. For example, a Community Farm in Chicago, B SP

Spock Nutritional Delivery Systems Vert Bhi Rack Designs, optimizing the use of resources in a limited urban space. This adaptation enables the farm to improve the quality of the yield while reducing waste and energy consumption. In both references, the normal thread is the empowerment of end users through local innovation and customization. AM allows the co-forming of tools and techniques to farmers, arbitrators and community organizations that fits their unique needs, eliminating traditional agricultural supply chains and standards of standard equipment (Bimpizas-Pinis et al., 2022).

Collaborative Platforms and Open-Source Design

One of the most promising trends in AM's application in agriculture is the rise of collaborative platforms and open-sun design communities. Platforms such as fireboats, an open-source accuracy farming project, provides digital blueprints and hardware specifications for 3D-printable farm automation tools. Users contribute and modify globally to make their local conditions better adapt, creating a dynamic ecosystem of innovation. Similarly, Thing-views host large libraries of agricultural equipment design, from seed dispensers to pest monitoring devices, enable users to download, adapt and print on demand. These platforms reduce the obstacles of technical access and global knowledge, promote the proliferation of sustainable farming solutions (Ciccullo et al., 2021).

The collaborative nature of this platform also promotes the engagement of multi-style, bringing farmers, engineers, designers and researchers together. These convergence supports the repetitive design cycle informed by real-world response, leading to constant improvement in agricultural technologies. Importantly, the open source is designed with the principles of the circular economy by promoting re -use, repair and local adaptation instead of production and disposal. Innovations in additive manufacturing are faster shaping the future of sustainable agriculture in various geographical and farming systems. In urban laboratories enables the use of Customization, waste reduction and circular resource, from cutting-age pilot projects to practical bottom solutions in rural communities. These benefits enhance these benefits by democratization of collaborative platforms and open-cell design technical access and promoting the global innovation community. This case study shows the AM's important role in driving elastic, efficient and incorporated agricultural practices associated with the principles of the circular economy.

CHALLENGES AND BARRIERS TO ADOPTION

While additive manufacturing (AM) promises significant promises to expand the sustainable agriculture in the circular economy, its widespread adoption faces many significant challenges. These obstacles include technical and structural limitations, skills and digital literacy gaps and environmental as well as economic considerations. A.M. in agriculture It is important to overcome these obstacles to understand the full potential of technologies (Ciccullo et al., 2021; Smith, 2025).

Technical and Infrastructure Constraints

One of the leading challenges for adoption of additive manufacturing for agricultural programs is the technical complexity and structural requirements. A high-quality 3D printer capable of producing durable, precision agricultural components can be expensive and requires stable power supply, maintenance and calibration. This is especially problematic in rural or remote farming regions where access to reliable electricity and technical support is limited. Moreover, the availability of proper printing material is a barrier. While bio-based and recycled filaments are emerging, they often have a lack of mechanical power or environmental durability for many agricultural equipment exposed under harsh conditions. Certificate of limited standardization and such content can reduce the user's trust in their life and safety. In addition, the integration of AMs with other technologies - such as IoT sensors, robotics and data analytics - requires civilized digital infrastructure and interoperability standards. Without strong connectivity and consistent Software ecosystems, the full benefits of smart, AM-enabled farming devices cannot be experienced.

Skills, Digital Literacy, and Access

There is a demand for skilled workforce capable of operating 3D printers for adoption of additive manufacturing in agriculture, conducting design Software and doing regular maintenance. Many farmer communities, especially in developing regions, face the shortage of such technical skills. Lack of Formal Training Programs or Access of Educational Resources Case hinders farmers from using AM techniques independently. Moreover, digital literacy hinders the effective use of garbage design platforms, open-sun repositories and digital fabrication workflow. Even when printer and design are accessible, users can struggle with file formats, cutting software ftware or printer troubleshooting, which causes an inefficient product or failed print. Access cess inequality also reflects widespread socio-economic division. High initial investment costs for printers and materials, with limited credit options, prohibits adoption in better resource farms or institutions. Small Holder Farmers - The spine of global agriculture is especially susceptible to excluding, endangering digital division that can increase inequality rather than promoting stability.

Environmental and Economic Considerations

Despite its environmental advantages, the durability of AM is profile and reference based. energy minimum consumption of 3D printers, especially industrial grade machines, can be significant. When powered by fossil fuel or inefficient grid, the carbon footprint of AM can offset some circular economy benefits. Lifelong assessments suggest that careful energy is required to improve sourcing and efficiency to maximize environmental benefits. On the material front, when bio-based filaments promise, their production processes and the final options of life are not fully PTIMES. Some bioplastics require industrial dysfunction facilities, which are not available in many rural areas, leading to potential landfill accumulation. On the contrary, the traditional plastic used in many AM filaments, worries about microplastic pollution if not properly operated (Smith, 2025).

Economically, when AM reduces costs related to inventory and transportation, initial capital costs and current operational costs may be restricted. Moreover, the quality of print, reliability of the part, and uncertainties around the certificate can disappoint invest in serious agricultural components. The acceptance of the market depends on the development of strong beliefs and standards to ensure a benchmark of traditional product quality or more. Adoption of additive manufacturing in sustainable agriculture is the technical, educational, environmental and economic barriers to each other. Integrated efforts in infrastructure development, capacity construction, content innovation and policy support are required to eliminate these challenges. Its advantages are required to ensure that AM technology and training are required to ensure that all farmers, especially small holders and marginalized communities. Honestly A.M. with the goals of the circular economy. It is important to address energy zones and material durability for aligning practices. Through exploring these obstacles, stakeholders can unpliant to the transformative prospect of AM to promote elastic, efficient and incorporated agricultural systems.

POLICY AND STRATEGIC FRAMEWORKS

Effective adoption and scaling of additive manufacturing (AM) in strong policy and strategic structure in models of sustainable agricultural and circular economy. Government, institutions, and stakeholders play a key role in promoting innovation, ensuring similar equitable access, and protecting environmental and safety standards. This department searches for major actors, investment and education initiatives, and development of standards and rules that consider sustainable AM deployment in agriculture (Melnyk et al., 2023).

Role of Government and Institutions

Governments and public institutions are crucial capable of additive manufacturing integration in agriculture. Their role extends beyond the funds to incorporate policy policies, structural development and ecosystem building (Ruzinazarov et al., 2024).

- Policy and incentives: Governments can compose policies that can promote the use of AMs for agricultural stability, such as subsidies for buying printers, tax incentives for using bio-based materials and AM in agriculture. Grants for research in the application. Policymakers should also include AMs in widespread agricultural modernization and circular economy strategy to ensure alignment and relevance.

- Infrastructure and support services: Public investment in digital infrastructure, such as reliable electricity, internet connectivity and fabric hubs (fablobs), especially in rural and underled areas. This hub can serve as community resource centers for AM technology, cess, maintenance and training.
- Collaboration and Innovation Network: Institutions such as universities, research centers and expansion services can promote multi-stakeholder cooperation in farmers, manufacturers, designers and technologists. Governments can facilitate innovation clusters and public-private partnerships that accelerate adaptation in technical transfer and local agriculture context.

Investment, Education, and Training Initiatives

To understand its potential in agriculture, human capital development and financial support are supreme(Smith, 2025).

- Investing in Capacity: Governments and development agencies should prioritize funds for technical training programs in view of farmers, agricultural and local technologies. Such applications must cover 3D printing operation, digital design skills, maintenance and troubleshooting to create self -reliance and reduce dependence on external specialists.
- Agricultural Education: Consolidating the concepts of AM technology and circular economy in professional schools, universities and expansion programs will be prepared to make future pay generations sustainable innovative. The platform and mobile learning can expand the training access to remote areas.
- Affordable financing. Success: Innovative credit methods such as microloans, leasing plans for 3D printers and public-private co-invested models can reduce access barriers for smaller holders and

cooperatives. Quality AM. The access to access to equipment and materials is crucial to adopt the same convenience.

Standards and Regulations for Sustainable AM

The development of clear standards and regulatory structures is important to ensure quality, safety and environmental stability in AM for agriculture(Giakoumatos & Tsoulfas, 2025; Tege et al., 2025).

- Material materials and production standards: Installing guidelines for acceptable materials, including bio-based and recycled filaments, ensures the parts are safe, durable and environmental sound. Certificate schemes can create user trusts and facilitate market acceptance of agricultural equipment and components produced by AM.
- Environment Environmental Compliance: Rules should order eco-lobbying for life cycle assessments and AM products to promote transparency of environmental effects. In AM processes, the policies promoting the use of renewable energy rooms and life-life-management (eg, recycling, composting) with the principles of the circular economy.
- Intellectual property and open innovation: Open-source design promotes innovation to balance the right of intellectual property with design framework while protecting manufacturers. Policy decisions must create frameworks that support collaborative innovation that hinders the spread of technology.
- Safety and Operational Guide: AM in agricultural settings: Safety, Operator Training and Maintenance. Reduces the risks associated with the use of devices, protecting both users and the end customers.

Policy and strategic framework are basic for scaling additive manufacturing as a means for sustainable agriculture in circular economy models. Governments and organizations can eliminate the current obstacles and promote the innovative AM ecosystem by applying support policies, investing in human capital and establishing rigorous standards. Strategic coordination between public and private sector, academia and farmer communities is required to ensure that AM technologies are accessible, environmentally durable and economically viable in various agricultural contexts.

Figure 5. Challenges and future developments in additive technology in sustainable agriculture for circular economy

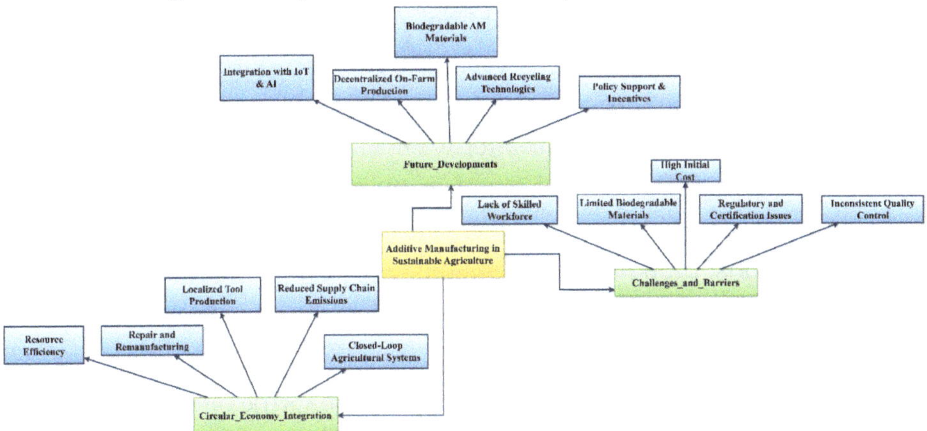

FUTURE DIRECTIONS AND OPPORTUNITIES

Additional Manufacturing (AM) is set to revolutionize the sustainable agriculture in the structure of the circular economy by developing technical integration, material innovation and scalable regenerative methods. This section detects emerging trends, highlighting the opportunities for ER Tanda synergy with AI, IoT and robotics, progress in biodegradable and circular materials, and a vision for a strong scalable, revived agricultural systems(Smith, 2025; Tege et al., 2025). Figure 5 highlights the challenges and future developments in additive technology for sustainable agriculture and the circular economy.

Integration with AI, IoT, and Robotics

The future of additive manufacturing in agriculture is inseparable with the convergence of artificial intelligence (AI), Internet Th F -Things (IoT) and robotics - technologies that enable collectively smarter, more efficient and adaptive farming solutions.

AI-powered design and optimization: AI algorithms can enhance AM processes by design parameters for durability, resource efficiency and efficiency of specific agricultural needs. Machine learning models can analyze consumption data to predict wear and promote continuous production evolution. IoT-enabled Smart Farming Devices: The integration of AM-produced ingredients with IoT sensor facilitates real-time monitoring of soil conditions, crop health and machinery performance. This synergy allows precise, data -based decisions and automation, reducing waste and improving yields(Kabato et al., 2025).

Robotics and Autonomous Systems: AM enables the production of custom parts for rapid prototyping and agricultural robots - crop monitoring, automatic irrigation systems or drones for robotic harvests. The modality and flexibility of 3D printing supports the rapid adaptation to the developed farm requirements and the atmosphere. Together, these technologies create a respondent and interconnected agricultural ecosystem where additive manufacturing supports agile hardware development and deployment.

Development of Biodegradable and Circular Materials

The innovation is crucial to enhance their environmental stability in agriculture(Giakoumatos & Tsoulfas, 2025).

Biodegradable Filaments: Future issues towards extensive adoption of biodegradable and compostable printing materials obtained from renewable resources such as polylectic acid (PLA), cellulose or starch-based polymer. This material reduces the dependence on fossil fuel and reduces environmental action when properly disposed.

Circular Material Loops: Progress in recycling techniques will enable the re -recovery and reuse of the waste and end parts of the end of life, closing the material loops. For example, farm on-perm recycling systems can process plastic parts used in new filament, producing material input costs and waste.

Functional and Bio-Composite Matcrials: Emerging bio-composites containing natural fibers or nanoparticles can increase mechanical properties while maintaining biodegradability. These innovations expand the scope of agricultural equipment produced by AM, making them stronger and durable.

Standardization and certification: In order to ensure future framework safety, display and original stability, emphasize bio-based and recycled materials, promoting extensive market acceptance.

Vision for Scalable, Regenerative Agriculture

Additional Manufacturing Scalable provides a way to regenerated agriculture that restores ecosystems when increasing productivity(Giakoumatos & Tsoulfas, 2025).

Local and On-Demand Product: AM facilitates decentralized production, reducing transportation emissions and supply chain weaknesses. Farmers can produce equipment and replacement parts corresponding to local references, empowering self -reliance and elasticity.

Customized Solutions for Resurrection: AM enables the formation of specialized equipment that supports regenerative methods-PCTs, clay aerodynamic equipment or water-efficient irrigation components or water-efficient promotes the health and biochemical of water.

Collaborative Innovation Networks: Future will see more open-source platforms and community-based design repositories where farmers, researchers and manufacturers co-create resurrected agricultural solutions enabled by AM.

Policy and Market Support: AM infrastructure corresponding to agricultural regeneration targets for scaling regenerated approaches requires sustainable methods and support policies promoting investment.

The future way of additive manufacturing in agriculture is shaped by AI, IoT and robotics, biodegradable and circular materials and shared vision for scalable regenerative farming. These opportunities are positioned not only as a manufacturing tool but also as the foundation of elastic, durable and circular agricultural system. Strategic collaboration, innovation and supporting policies will be important for unclosing this potential, eventually contributing to global food security and environmental stability.

CONCLUSION

Additional manufacturing circular presents a transformative probability to further sustainable agriculture in the structure of the economy. By enabling the local, custom product of equipment, ingredients and smart farming devices, AM resources address crucial challenges such as efficiency, waste reduction and supply chain elasticity. Its integration with emerging technologies such as AI, IoT, and robotics increases more precision and adaptability, paving the way for innovative farming methods. However, in order to realize this potential, efforts are needed to integrate with a support policy framework, investments in infrastructure and skills development, and biodegradable, adoption of circular content. As shown by global case teaching and pilot projects, AMs can stimulate the transfer towards regenerated agriculture-global, environmentally friendly and economically viable.

Ultimately, additive manufacturing is more elastic, durable agriculture as a major capable in the future transition, which contributes significantly to food security and environmental stewards worldwide.

REFERENCES

Balla, A., Silini, A., Cherif-Silini, H., Chenari Bouket, A., Alenezi, F. N., & Belbahri, L. (2022). Recent advances in encapsulation techniques of plant growth-promoting microorganisms and their prospects in the sustainable agriculture. *Applied Sciences (Basel, Switzerland)*, *12*(18), 9020. DOI: 10.3390/app12189020

Bimpizas-Pinis, M., Santagata, R., Kaiser, S., Liu, Y., & Lyu, Y. (2022). Additives in the food supply chain: Environmental assessment and circular economy implications. *Environmental and Sustainability Indicators*, *14*, 100172. DOI: 10.1016/j.indic.2022.100172

Chowdhury, H. (2023). Circular economy integration in additive manufacturing. *Preprints, Oct, 3.*

Ciccullo, F., Cagliano, R., Bartezzaghi, G., & Perego, A. (2021). Implementing the circular economy paradigm in the agri-food supply chain: The role of food waste prevention technologies. *Resources, Conservation and Recycling*, *164*, 105114. DOI: 10.1016/j.resconrec.2020.105114

Fu, X., Zheng, Z., Sha, Z., Cao, H., Yuan, Q., Yu, H., & Li, Q. (2022). Biorefining waste into nanobiotechnologies can revolutionize sustainable agriculture. *Trends in Biotechnology*, *40*(12), 1503–1518. DOI: 10.1016/j.tibtech.2022.09.013 PMID: 36270903

Giakoumatos, S. D., & Tsoulfas, G. T. (2025). Towards Shaping Integrated Circular Economy Approaches in Animal Breeding Supply Chains by Mainstreaming Microalgae Utilization. *Circular Economy and Sustainability*, *5*(7), 1–41. DOI: 10.1007/s43615-025-00520-8

Hettiarachchi, B. D., Brandenburg, M., & Seuring, S. (2022). Connecting additive manufacturing to circular economy implementation strategies: Links, contingencies and causal loops. *International Journal of Production Economics*, *246*, 108414. DOI: 10.1016/j.ijpe.2022.108414

Kabato, W., Getnet, G. T., Sinore, T., Nemeth, A., & Molnár, Z. (2025). Towards climate-smart agriculture: Strategies for sustainable agricultural production, food security, and greenhouse gas reduction. *Agronomy (Basel), 15*(3), 565. DOI: 10.3390/agronomy15030565

Khan, F., Pandey, P., & Upadhyay, T. K. (2022). Applications of nanotechnology-based agrochemicals in food security and sustainable agriculture: An overview. *Agriculture, 12*(10), 1672. DOI: 10.3390/agriculture12101672

Lu, Y., Xu, W., Leng, J., Liu, X., Xu, H., Ding, H., Zhou, J., & Cui, L. (2024). Review and Research Prospects on Additive Manufacturing Technology for Agricultural Manufacturing. *Agriculture, 14*(8), 1207. DOI: 10.3390/agriculture14081207

Malik, L., Sanaullah, M., Mahmood, F., Hussain, S., Siddique, M. H., Anwar, F., & Shahzad, T. (2022). Unlocking the potential of co-applied biochar and plant growth-promoting rhizobacteria (PGPR) for sustainable agriculture under stress conditions. *Chemical and Biological Technologies in Agriculture, 9*(1), 58. DOI: 10.1186/s40538-022-00327-x PMID: 37520585

Melnyk, L., Kalinichenko, L., Kubatko, O., Dobrowolski, Z., & Babczuk, A. (2023). *Restructuring the economic systems on the way to an additive economy.*

Parmar, P., Kumar, R., Neha, Y., & Srivatsan, V. (2023). Microalgae as next generation plant growth additives: Functions, applications, challenges and circular bioeconomy based solutions. *Frontiers in Plant Science, 14*, 1073546. DOI: 10.3389/fpls.2023.1073546 PMID: 37063190

Ruzinazarov, S., Achilova, L., Ismailova, R., & Men, V. (2024). Digitalization and legislative support in intellectual property protection for fodder additive production technologies. *BIO Web of Conferences, 149*, 01010.

Smith, A. (2025). Systemic Analysis of Circular Economy Principles in Agricultural Waste Management and Their Synergy with National Sustainable Development Objectives. *Transdisciplinary Advances in Social Computing, Complex Dynamics, and Computational Creativity, 15*(2), 10–21.

Sohail, M., Pirzada, T., Opperman, C. H., & Khan, S. A. (2022). Recent advances in seed coating technologies: Transitioning toward sustainable agriculture. *Green Chemistry*, *24*(16), 6052–6085. DOI: 10.1039/D2GC02389J

Soni, R., Gupta, R., Agarwal, P., & Mishra, R. (2022). Organic farming: A sustainable agricultural practice. *Vantage: Journal of Thematic Analysis*, *3*(1), 21–44. DOI: 10.52253/vjta.2022.v03i01.03

Tege, S., Jangeed, D., & Soni, B. (2025). Exploring the role of digital technologies in promoting circular economy practices in the Indian agriculture sector. In *Water Sustainability and Hydrological Extremes* (pp. 333–357). Elsevier. DOI: 10.1016/B978-0-443-21499-8.00017-9

Xu, Q., Zhang, T., Niu, Y., Mukherjee, S., Abou-Elwafa, S. F., Nguyen, N. S. H., Al Aboud, N. M., Wang, Y., Pu, M., Zhang, Y., Tran, H. T., Almazroui, M., Hooda, P. S., Bolan, N. S., Rinklebe, J., & Shaheen, S. M. (2024). A comprehensive review on agricultural waste utilization through sustainable conversion techniques, with a focus on the additives effect on the fate of phosphorus and toxic elements during composting process. *The Science of the Total Environment*, *942*, 173567. DOI: 10.1016/j.scitotenv.2024.173567 PMID: 38848918

Chapter 8
Recent Innovations and Emerging Trends on Genetic Engineering and Biotechnology in Agronomy

B. M. Parashiva Murthy
https://orcid.org/0000-0001-8808-8366
Department of Computer Science and Engineering, JSS Science and Technology University, Mysuru, India

Rajesh V. Argiddi
https://orcid.org/0000-0002-4227-2384
Department of Computer Science and Engineering, Walchand Institute of Technology, Solapur, India

Gunti Bhupal Raj
https://orcid.org/0009-0009-5618-2388
School of Agriculture, SR University, Warangal, India

M. Prasanth
https://orcid.org/0009-0005-4617-9805
Nandha Engineering College, Perundurai, India

Venkateswara Rao Podile
https://orcid.org/0000-0001-5251-8424
KL Business School, Koneru Lakshmaih Education Foundation, Vaddeswaram, India

M. Suresh
https://orcid.org/0009-0006-5764-1660
KPR Institute of Technology, Kovai, India

DOI: 10.4018/979-8-3693-4392-0.ch008

ABSTRACT

Recent progresses in genetic engineering and biotechnology are revolutionizing crop yields, resistance to pests and diseases and environmental stability. Techniques such as CRISPR-KAS9, transgenic modification and molecular marker-support selection are enabling specific feature improvement and rapid breeding cycle. These innovations contribute to climate-elastic agriculture, reduction of agrochemical use and improved crop nutrient profiles. Moreover, biotechnological programs are supporting land health and sustainable farming practices. This chapter detects current progresses, practical applications and future possibilities in agronomies that highlight their transformative impact on global food security and agricultural productivity.

INTRODUCTION

The agricultural system worldwide is under unprecedented pressure due to weather reversal, population growth, reducing natural resources and rising food protection. Traditional farming techniques, despite the basic for modern agriculture, are often inadequate to consider the scale and complexity of today's challenges. In this regard, genetic engineering and biotechnology have emerged as the main forces resurfacing the landscape of agronomy. This encyclopedia has opened transformative possibilities to improve the elasticity of the crop, increase the potential for yield and ensure environmental durability. Biotechnology, in its broadest sense, uses living organisms, systems or biological processes to develop agricultural products and processes, while genetic engineering especially includes direct trafficking of the genome of the organism using nuclear equipment. Together, they create a strong toolkit to overcome the multifaceted challenges faced by modern agriculture (Demirer et al., 2021a).

One of the most significant progresses in this area is the development and application of genome editing techniques. Compared to the previous genetic modification methods, CRISPR-KAS9 is less time consuming, more cost-effective and significantly more accurate, reducing the risk of unnecessary genetic consequences. This innovation has accelerated the development of climate-smart crop varieties, which are increasingly necessary to reduce the influence of irregular weather patterns, land degeneration and water scarcity. Moreover, transgenic technology continues to play an important role in the development of insect-resistant and herbicide-tolerant crops, such as BT cotton and glyphosate-resistant soybeans, who have taken significant agricultural and economic influences in countries such as India, Brazil and the United States(Gao, 2021).

Another crucial field of development is the marker-help selection (MAS) and genomic selection, which combines classical breeding methods with nuclear living. MAS has been especially useful in improving complex symptoms such as yield stability, immune resistance and abysiotic stress tolerance, which is influenced by multiple genes and environmental interaction. In addition, plant tissue culture and micropropagation techniques have enabled a large amount of disease-free, high-quality planting material, especially for horticultural and planting crops, there is improvement in productivity and profitability for farmers (Mamata & Lamichhane, 2021; Upadhyay, 2021).

Biotechnology also contributes significantly to sustainable agriculture. Genetically engineer crops that require minimum chemical inputs reduce the ecological measures of farming. For example, crops with built-in resistance in pests can reduce the need for pesticide application, preserve beneficial insects and reduce groundwater contamination. Similarly, nitrogen-efficient genetically improved crops can PT on the use of fertilizer, there are significant contributors, emissions and nutrient part-regulatory degeneration. Moreover, the development of biopeptides originating from beneficial microorganisms offers an environmental -friendly option for artificial agrochemicals, promoting soil health and promoting integrated paste management (IPM) methods(Hamdan et al., 2022).

From a socio-economic point of view, biotechnological innovations are likely to improve the livelihood of small holder farmers by increasing crop productivity, reducing loss and enabling high-value crops. In developing countries, where agriculture is both livelihood and method of existence, genetically improved seed access cess can mean the difference between subsistence and food security. However, the adoption of genetically modified (GM) crops remains uneven in regions due to regulatory, moral and public assumption challenges. Countries such as the United States, Canada and

Argentina have accepted GM crops, while other people have maintained moratoria or strict regulatory structures due to biosfty concerns, intellectual property rights issues and corporate control of seed systems. These concerns highlight the importance of transparent regulation, participating technical development and incorporating innovation framework that consider the interests of all stakeholders, including farmers, consumers, entrepreneurs (Kalendar et al., 2022; Yan et al., 2022).

In parallel, the conversion of digital agriculture with biotechnology is new opportunities for precision agronomy. The integration of genomics with large data analytics, artificial intelligence and remote sensing enables more informed decisions at the farm level. For example, genomic data can be used to predict plant operations in specific environmental conditions, which allow the strategic deployment of varieties in the targeted agro-parable zone. Moreover, bioinformatics tools simplify the identity of gene-defense organizations and accelerate the detection of candidate genes for further improvement. Such development points to the future in which genetic engineering is not a different activity, but an integral component of the large, data -based agricultural innovation ecosystem(Squire et al., 2023a).

Despite the optimism around this technological advancement, it is necessary to contact the future of biotechnology in agriculture with a balanced perspective. Genetic manipulation, biodiversity protection and moral consideration around the same access to technology should be at the forefront of the ethnic and policy dialogues. The precautionary principle associated with strict risk assessments, especially in ecological sensitive or socio-economically sensitive regions, should guide the deployment of new biotechnologies. Public connection and education are also crucial in rejecting genetic engineering, eliminating misconceptions, and enhancing confidence in the innovations based. Ultimately, the success of biotechnology in Agronomy will depend not only on the excellence(Lallawmkimi et al., 2024; Slafer et al., 2024).

Hence, Genetic engineering and biotechnology represent an instance shift in agronomy, providing innovative solutions to some of our time's most agricultural challenges. From CRISPR-based genome acquisition to molecular breeding and from microbial bio-input to digital bioinformatics platform, these technologies are operating a new era of intelligent, durable and elastic farming. This chapter will look for recent innovations in the field, showcase practical programs, and discuss future prospects and policy considerations that will shape the path of agricultural biotechnology(Ali et al., 2025; Stewart Jr, 2025a). As we explore the difficulties of global food systems in the 21st century, traditional agriculture. Closing the cutting edge of consolidation with wisdom will be the key to cultivating a sustainable and food-shaking future.

Objectives

The purpose of this chapter is to explore the latest progress of genetic engineering and biotechnology applied to Agronomy. The main objectives include investigations into CRISPR-KAS9, transgenic crop development, and marker-support selection and modern technologies such as their roles in improving crop productivity, elasticity and durability. It also tries to analyze the integration of biotechnological tools in precision agriculture and evaluate their socio-economic and environmental effects. Moreover, this chapter will affect regulatory structures, moral considerations and the influence of adopting these techniques. After all, it aims to provide a strategic view on how these innovations can shape future agronomic practices.

BACKGROUND GENETIC ENGINEERING AND AGRICULTURAL BIOTECHNOLOGY

Genetic engineering and agricultural biotechnology represent the basic pillar of the modern agronomic vision, which gives a transformative possibility in eliminating crucial global challenges related to food security, weather change and environmental stability. At their origin, genetic engineering refers to the deliberate change of the genome of the organism through the direct manipulation of its DNA, often to give the desired traits, change the special genes. Agricultural biotechnology includes a wide range of enemy equipment and techniques used to improve the productivity, quality and elasticity of a comprehensive domain, crop, livestock and other farming systems. In this field, integral key concepts include recombinant DNA technology, gene cloning, transgenesis, genome editing and molecular diagnostics (Tyczewska et al., 2023). While genetic engineering focuses on cellular and nuclear levels to enhance or represent symptoms such as drought resistance, insect tolerance or nutrient efficiency, biotechnology extends to practical programs, including accuracy, disease testing, biofertilizers and durable pest control. These branches are integrated by the purpose of improving the efficiency and durability of agricultural production while reducing the ecological footprint and input intensity of traditional methods.

Historical evolution of genetic engineering and biotechnology in agriculture is marked by a range of qualified progress and regulatory goals. The foundation of biotechnology was placed in the beginning of the 20th century with a re -detection of the laws of Gregore Mendel's inheritance, which depended for modern genetics(Tyczewska et al., 2023). However, the actual change began with the advent of recombinant DNA technology in the 1970s, enabling one organism to enter genes from another. In 1983, the first genetically modified (GM) plant, antibiotic-resistant tobacco was produced, which jumped significantly in applicable agricultural biotechnology. The most transformed landmark in recent years is the development of CRISPR-Casa genome editing technology in 2012, enabling the specific, efficient and cost-effective acquisition of the plant genome. This innovation has opened new ways for non-transgender genetic improvement, making the lines between traditional breeding and modern biotechnology more obscure. As agricultural research became increasingly interdependent, the wave of integrated innovations was created in collaboration between geneticists, nuclear organisms, ologists, bainites and agriculture (Demirer et al., 2021b).

Different range of tools and techniques regarding the current landscape of genetic engineering and agricultural biotechnology. The most crucial tools include recombinant DNA technology, including combining DNA sequences from different sources to create new genetic compounds. This is necessary to develop transgenic crops that carry genes for desired symptoms such as insect resistance or herbicide tolerance. It enabled to distinguish and mimic the specific genes of interest, while DNA sequencing allows the full mapping of agionym - agriculture is essential to identify the genes associated with the vital symptoms(Abdul Aziz et al., 2022; Demirer et al., 2021b).

Polymerase chain reaction (PCR) is widely used to expand DNA segments and detect the presence of special genetic elements, which is a crucial step in both genetic modification and diagnostic applications. Molecular markers, such as SSR (Simple Sequence Repeats) and SNP. (Single nucleotide polymorphism) is important for marker-supporting selection, facilitating faster and more accurate breeding programs by identifying plants that carry desirable symptoms at the plant stage. Tissue culture and micropropagation techniques are used to quickly multiply the same and disease -free planting material (Squire et al., 2023b; Verma et al., 2022). They also serve as a platform for genetic mutations, in which foreign DNA plant cells are introduced to the use of techniques such as agrobacterium-medial transformation. In recent years, CRISPR-KAS9 and relevant genome editing tools have been famous for their accuracy and ease of targeting targeted bypassing the regulatory barriers associated with transgenic crops, without introducing foreign DNA. Moreover, bioinformatics tools and genomic databases play a crucial role in storing, analyzing and interpreting complex genetic information, facilitating the identity of gene-construction organizations and enabling prediction breeding models.

Together, these concepts, historical hymn progresses and technological advances create the foundation of modern agricultural biotechnology. The integration of their agricultural into has significantly benefited the yield, quality, stress tolerance and input efficiency of the crop. As biotechnology continues to develop, its role in shaping the future of agriculture will only enlist through responsible research, innovation and commitment to durability.

INNOVATIONS IN GENETIC ENGINEERING

CRISPR-Cas and Genome Editing Technologies

Innovations in genetic engineering have revolutionized modern agriculture. The revolution in the landscape of the land, which enters the age of the possibility of unprecedented precision, efficiency and durable crop improvement. The most transformative progress is the development of genome acquisition techniques, especially CRISPR-CAS (clustered regularly interconnected short palindromic repetitions). CRISPR-CAS technology allows the edit of plant genomes with extraordinary specificity by targeting special DNA sequences and inducing cuts to certain genomic places (Ali et al., 2025; Lallawmkimi et al., 2024).

Figure 1. Innovations in genetic engineering have revolutionized modern agriculture

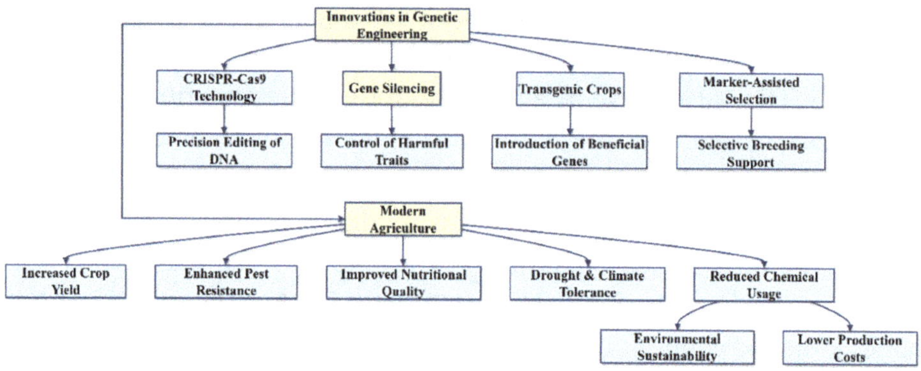

This facilitates, eliminating, inserting or modifying genetic material with minimal -reflection effects, enables the development of advanced symptoms such as immune resistance, Abiotic stress tolerance and improved nutrient value. Unlike conventional transgenic methods, which are often inserted to foreign DNA in the host genome, the CRISPR enables a specific gene change without the introduction of external genes - this possible regulatory barrier and public resistance associated with improved organisms (GMOs). CRISPR's requests in Agronomy are wide, including the development of drought resistant rice, immune-resistant wheat and non-potato mushrooms. In rice, for example, CRISPR is used to harden the genes associated with yield penalties under drought stress, resulting in high productivity varieties in water-limited environments(Gao, 2021; Tyczewska et al., 2023). In tomato and soybeans, fungal pathogens have tolerated without compromising the characteristics of gene acquisition. With technological advancement, new types of base editing and prime editing are enhancing the accuracy and versatility of genome changes, expanding the possibilities for feature stacking and multiplex acquisition, where many genes are simultaneously edited for complex feature growth. Figure 1 shows that innovations in genetic engineering have revolutionized modern agriculture.

Transgenic Crop Development

Supplementing genome acquisition is the continuous progress of transgenic crop development, which includes the introduction of foreign genes in plant genomes to provide new features that are not present in the original gene pool. After the first GM crop commercialization in the 1990s, transgenic technology has significantly improved global agricultural productivity. BT crops (eg, BT cotton, BT maize), which contain bacterium bacillus thinginess genes that produce toxic pesticide protein in specific pests(Lallawmkimi et al., 2024; Slafer et al., 2024).

This crop has contributed to chemical pesticides, improvement in yield stability and expansion of durable pest management systems. Similarly, herbicide-tolerant crops, such as glyphosate-resistant soybeans and corn, allow farmers to operate less farming and reducing herbicide inputs. In recent years, transgenic approaches have been used to enhance crop quality symptoms-as a result, golden rice, engineer to produce beta-carotene, addresses vitamin A deficiency in developing countries. The second example contains high oleic acid soybean oils, developed to provide better oxidative stability for heart-health benefits and cooking. With the ongoing purification of transformation methods, transgenic technology now allows more precise gene insertion and expression regulation, reducing the risk of gene silence or unwanted effects. In addition, the tissue-specific and inducible promoters enable a gene expression controlled in specific plant parts or in certain environmental conditions, improving both safety and efficiency, in the front or in certain environmental conditions. Despite its enzyme. Despite its success, transgenic crops are adopted, including regulatory sanctions, intellectual property discussions and consumer skepticism. More public engagement, transparent safety assessments and innovation policies are needed to ensure the same access cess and acceptance to eliminate these challenges.

Synthetic Biology in Agronomy

In parallel, the emerging field of artificial organisms is expanding the boundaries of genetic engineering in Agronomy by the creation and construction of new biological systems or redesigning existing issues for useful agricultural works. Artificial organisms combine the principles of engineering, computational biology and molecular genetics that create standard genetic parts and circuits that can be presented in plants to perform specific, predictable tasks. A key application is the formation of artificial metabolic pathways that enables plant novel compounds, such as biopesticides, pharmaceuticals or advanced micro -nutrients(Ali et al., 2025; Slafer et al., 2024).

For example, artificial organisms. An plants enable the biosynthesis of valuable secondary metabolism. In other innovative applications, researchers have engineered artificial nitrogen-fixing routes in non-leguminous crops such as corn and wheat, potentially reducing the need for chemical fertilizers and reducing associated environmental damage. Artificial gene circuits are also being developed to control the respondents of external stimuli plants, allowing the crop to be dynamically adapted to environmental conditions such as drought, salinity or pathogenic presence.

Additionally, Artificial promoters and transcript regulators provide fine-tune control over the gene expression, increasing the reliability of complex genetic changes. One more future vision includes plant biosensors - usually detected by visible symptoms or biochemical markers of earthen nutrients, pollutants or pest pressure. These innovations not only improve crop operations but also integrate with precision agricultural systems. Artificial organisms with automation and AI-driven design tools are accelerating the development pipeline, enabling fast prototyping and testing of engineered features.

In summary, CRISPR-Casa Genome acquisition, transgenic crop development and artificial life. No innovation in the agronomy is rebuilding genetic engineering landscape. Together, they offer a synergistic probability for making the next pay -generation crop that corresponds to the needs of more elastic, productive and future food systems. As these techniques mature, digital platforms, large data analytics and systems are expected to be connected with biology to create a holistic, intelligent structure for agricultural innovation. While challenges are in terms of regulation, morality and public perception, the enlightenment of these innovations is undeniable. Strategic investments, incorporated regime and responsible innovation will be required for the full benefits of genetic engineering to ensure durable and food-safe future.

Figure 2. Biotechnological applications in crop improvement

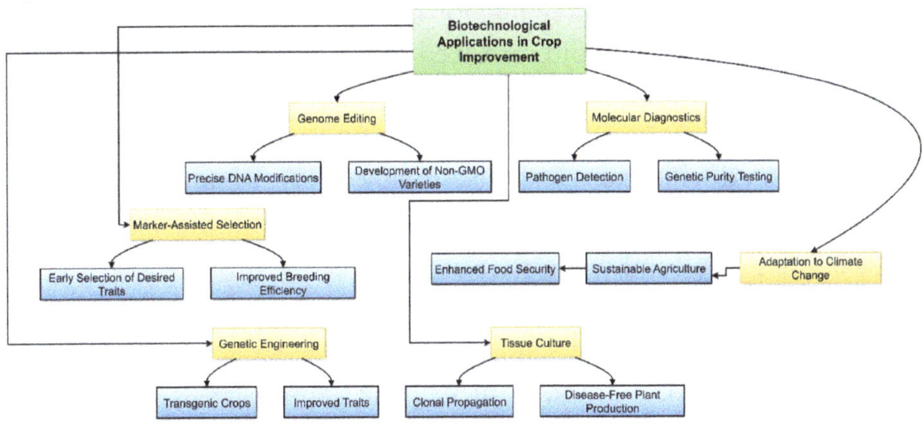

BIOTECHNOLOGICAL APPLICATIONS IN CROP IMPROVEMENT

Crop improvement through biotechnology has become the cornerstone of modern agricultural innovation, providing solutions to some of the most challenges faced by farmers and food systems worldwide. As climate change increases, pest pressure increases, and nutritional deficiencies persist in many areas, the strategic application of biotechnology is enabling the growth of crop varieties that are more flexible, productive and nutritional valuable. The most impressive applications have an increase in bi-fuel efforts with the aim of growing drought and stress tolerance, growth and immunity crops, and improving nutritional quality of staple foods. These progresses are not only addressing yield stability and food security, but are also contributing to more durable and resource-skilled agricultural systems. Figure 2 illustrates the various biotechnological applications in crop improvement(Ali et al., 2024; Zhang et al., 2025).

Drought and Stress Tolerance

Drought and abiotic stress represent some of the most immediate goals for crop improvement in view of the increasing frequency and intensity of tolerance, water scarcity, salinity, extreme temperature and other climate-induced stress. Traditional breeding for stress tolerance is often limited by the complexity of the symptoms involved, which is usually multilateral and influenced by environmental interaction. Biotechnology, however, allows more targeted approaches by identifying and manipulation of specific genes associated with stress responses(Kang, 2024; Zhang et al., 2025).

For example, transcript factors such as drab (dehydration response element binding), NAC and WRKY are used to increase drought tolerance by modulating a stress-responsive gene network. The transgenic crop showing these genes has shown the efficiency of water use, reduced stomatal conductivity and better osmotic adjustment in the state of stress. Additionally, biotechnologists have introduced genes that encode for protective proteins such as late embryogenesis (LEA) protein and heat-shock protein (HSP), which helps maintain cellular integrity during periods of abiotic stress.

In rice, wheat and corn, genetically modified lines have shown improved operations under drought and heat tension without significant yields in normal conditions. Biotechnology is also helping the development of crops that suffer the growing problem in the salinity - irrigation regions of the soil by the engineering sodium transporter genes and enzymes involved in the osmo-protectant synthesis. As climate elasticity becomes a priority, the role of biotechnology in the harsh environment will be important to sustain the role of biotechnology.

Pest and Disease Resistance

Another major domain of biotechnological application is insect and immune resistance, which directly affects crop damage, input costs and food quality. One of the most successful examples of this is the development of BT crops, which expresses pesticide protein obtained from Bacillus thuringiensis.

Additionally, RNA Intervention (RNAI) technology emerges as a powerful tool for silence genes in insect pests or pathogens, providing species-specific protection without the use of toxic chemicals. For example, RNAI based corn has been developed to deal with the rootworm of Western corn, which is the main insect of North America. Similarly, efforts are being made to make late -blooded resistant potatoes by incorporating genes from the wild Solanum species. Advanced gene acquisition equipment is now being used to remove sensitivity genes or enhance congenital immune system in the crop, enabling the development of broad-spectrum and sustainable resistance with minimal genetic disruption. These innovations not only reduce dependence on chemical pesticides, but also contribute to environmental stability and food safety(Bharat et al., 2024; Kang, 2024).

Nutritional Enhancement and Biofortification

In parallel, biotechnology is addressing the global challenge of nutritional growth and biofortification, malnutrition of microorganisms, often called "hidden appetite". Biofortification involves increasing the concentration of essential vitamins and minerals in the main crop, which makes nutrient -rich foods accessible to the population that depends greatly on grains and tubers for daily calorie intake. One of the most recognized successes in this area is the Golden Rice, engineered to produce Provitamin A (beta-carotene) in endosperm. This intervention is the main cause of prevention blindness and mortality in Asia and Africa children aiming for vitamin deficiency. Other biofortified crops include iron-rich rice and pearl millet, zinc-fortified wheat and advanced vitamin E and protein content. Genetic engineering also allows the synthesis of nutrients of the novel or the improvement of amino acid profiles in the main crop. Nutritional growth by biotechnology provides a long-term, food-based strategy to fight malnutrition, especially in resource-elegant settings where complement and diet variations are not always possible(Adhikari, 2024; Chandra et al., 2024).

Thus, use of biotechnology in crop improvement - drought and stress tolerance, pest and immunity and nutrient biofortification - Agriculture to meet the demand for global food and nutrition. These innovations enable more efficient, elastic and health -centered agricultural systems that adjust with sustainable development goals. In order to ensure the benefits of agricultural biotechnology, especially in developing regions where the impact can be the most transformative, proceeding to research, support regulatory structures and transmission strategies will be required, auxiliary regulatory strategies will be required.

Figure 3. Molecular breeding and Marker-Assisted Selection (MAS) function as powerful tools in modern crop improvement

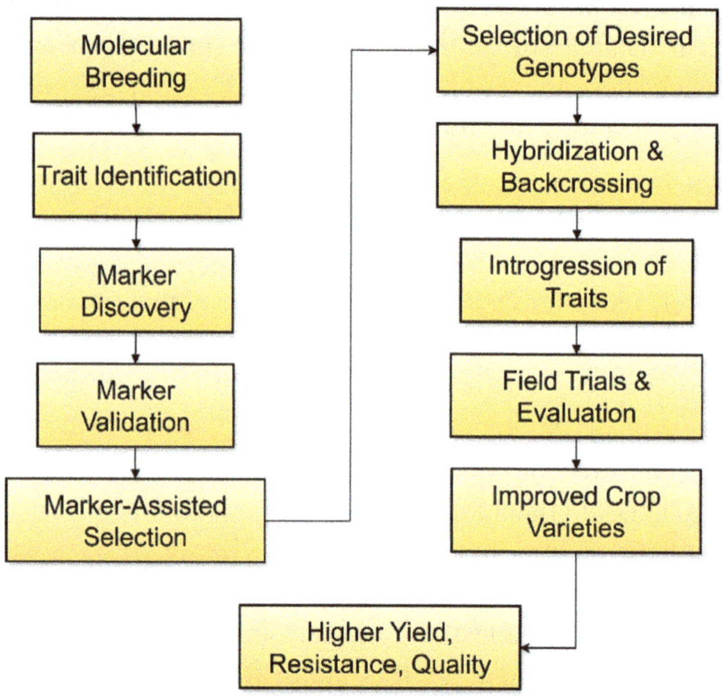

MOLECULAR BREEDING AND MARKER-ASSISTED SELECTION

Techniques and Strategies

Molecular breeding and marker-support selection (MAS) emerged as a powerful strategy in modern crop improvement, which enables breeders to make informed decisions on the basis of genotypic information rather than just depend on phenotypic symptoms. These biotechnological approaches increase the accuracy, motion and efficiency of breeding programs by identifying, tracking and selection of genes or genomic regions associated with symptoms. Atomic breeding is the use of nuclear markers - special DNA sequences that are tightly linked to target genes or quantitative feature Loki (QTL). These markers serve as a proxy for the symptoms of juice and can be used to select plants with allely at the initial growth stage, even before the symptom is phenotypically expressed(Ali et al., 2024; Zhang et al., 2025). Figure 3 illustrated how Molecular Breeding and Marker-Assisted Selection (MAS) function as powerful tools in modern crop improvement.

The application of MAS significantly reduces the breeding cycle time and cost, increases the probability of success in the choice of complex feature, and facilitates the pyramiding of multiple symptoms in the elite varieties. Ban Restriction Fragment Length Polymorphism, Simple Sequence Repeats (SSR), Single Nucleotide Polymorphism (SNP), Amplified Fragment Length Polymorphism (AFLP), and Genotyping-Genotyping (GBS) Technologies such as genotyping-by-sequencing (GBS). In this, SNP Markers have been most commonly used in recent years due to their abundance, high-thru-screening potential and Automation Reflection. Nuclear breeding quantitative feature also integrates mapping, association mapping and genomic selection (GS), where the breeding values of individuals are predicted on the basis of genome-wide marker data. These strategies allow for complex, multilateral symptoms, such as yield, stress tolerance and immune resistance, which enables more targeted and effective breeding efforts.

Case Studies and Success Stories

Stories of cases teaching and success show the transformative effect of nuclear breeding and MAS on global agriculture. The development and release of sub 1 rice varieties in rice, southern and southeast Asia is a landmark success. Sub 1 gene, tolerance to drowning stress, was identified in FR 13A variations of traditional Indian rice and entered high-yielding varieties like Swarna and IR 64 using marker-supported backcrossing (MABC). These varieties can survive for two weeks, significantly improve yields in flooded regions, and benefit millions of small farmers. Similarly, MA in-progress bacterial resistance in rice operates in elite planting in genes (XA21, XA13, XA5), resulting in resistance to durable, broad-spectrum immunity. In wheat, molecular markers have facilitated the development of rust-resistant varieties by the inclusion of genes like LR34, SR2 and YR 15, which protects against developed pathogen races(-Demirer et al., 2021b; Verma et al., 2022). Another significant example is found in maize, where the MASA has accelerated the development of drought-relevant hybrids under the initiative of drought tolerant corn for Africa (DTMA). This hybrid sub -Saharan has improved food protection and resilience for millions of people in Africa. League like chickpeas and pigeons, drought tolerance, markers attached to QTL for fusarium wilt resistance, and flowering time are used for stronger and productive planting. India's "hope" combines the possibility of an improved yield in the elasticity of drought, developed using various chickpeas, MAS. In horticultural crops like tomato, resistance to tomato yellow leaves curl virus has been mapped and deployed by MAS to stabilize the production in the affected regions. Collectively, these successes show the ability of MA to improve the elasticity of stress, disease tolerance, yield potential and quality symptoms in a wide range of crops.

Role in Accelerating Crop Breeding Cycles

The main benefit of nuclear breeding and MAS lies in their role to accelerate the breeding cycle of the crop. Traditional breeding generally requires many years and pay -generations that are required to identify and stabilize desirable symptoms through successive field tests and phenotypic evaluation. However, MAS enables plant selection based on their DNA profile, often at the stage of the seedlings, there is a lot of pay -generations of field testing. This initial and accurate choice shortens the breeding cycle and increases the throughput of breeding applications. For example, the rapid infiltration of one or a few genes of the donor while reducing the carrying of unwanted donor symptoms allows the marker-cooperative backcrossing allows the elite clever(Ali et al., 2024; Kang, 2024; Zhang et al., 2025).

This method reduces the number of backcrosses pay generations, especially from six to three, without compromising the re -recovery procurement of the genetic background. Moreover, the marker-helped recurrent selection (MARS) and Genomic Selection (GS) increase the enhancement efficiency by enabling simultaneous correction of multiple complex features powered by numerous small-edged genes. GS, in particular, uses numerical models to predict individuals based on genome-wide marker data, even in the absence of phenotypic evaluation. This approach is especially valuable for symptoms with low inheritance or for expensive and time consuming, such as root architecture or nutrient use efficiency. Integrating MA and GS with speed breeding techniques, which includes faster generation turnover under a controlled environment, compressing more breeding timelines and accelerating the release of improved varieties. The result is a more crisp and responsive breeding system that can eliminate emerging challenges such as climate change, developed insect pressure and changing market demand.

Hence, molecular breeding and marker-supporting selection represent modern agriculture. In the progress of the game-lit-ups. By giving the benefit of genetic information, these approaches enhance the accuracy, motion and space of crop reform programs. Molecular breeding is contributing to the formation of upcoming pay -generation, in line with the demand for growing and dynamic global food system, for the development and quality development of immunity from immune resistance and abiotic stress tolerance.

As genotyping costs decrease and genomic technologies become more accessible, expect to adopt MAS and genomic selection, especially in developing countries where the potential impact on productivity and elasticity is the highest. To fully realize the benefits of nuclear breeding, however, integration with capacity construction, investment in genotyping infrastructure and traditional breeding knowledge is essential. Ultimately, the synchronization between biotechnology and traditional plant breeding has the key to creating climate-elastic, high performance and peasant-preferred crop varieties for the future.

MICROBIAL AND SOIL BIOTECHNOLOGY

Biofertilizers and Biopesticides

Microbial and clay biotechnology emerged as a crucial component of sustainable agricultural development, offering alternatives to environmental fertility for chemical-intensive methods while enhancing soil fertility, crop productivity and ecosystem health. This intercourse field gives beneficial microorganisms - bacteria, fungi, actinomycetes and algae - to improve nutrient cycling, suppress pests and diseases, and restore degeneration (Singh et al., 2024). The most prominent applications include biofertilizers and biopesticides, which use live organisms or their metabolic products to perform agronomically beneficial tasks. Figure 4 illustrates the role of microbial and clay biotechnology in sustainable agriculture.

Figure 4. Microbial and clay biotechnology in sustainable agriculture

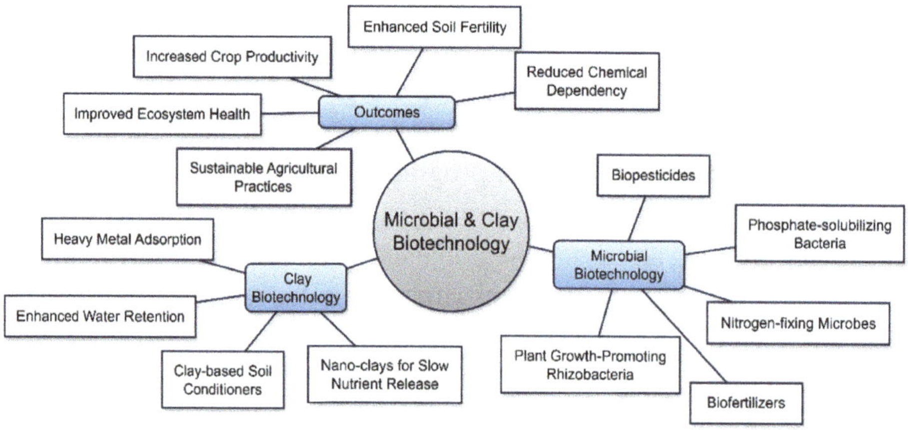

Biofertilizers, such as Rhizobium, Azotobacter, Phosphate Solubilizing Bacteria, and mycorrhizal fungi, biological nitrogen fixation, phosphorus solubilization and micronutrient dynamics. These inputs reduce the dependence on artificial fertilizers, low cost of production, and long -term soil health. For example, the rhizobium forms a symbiotic relationship with the leguminous crops, fixes the atmospheric nitrogen in the root nodules, and provides the renewable source of this essential nutrients (Zhang et al., 2025).

Similarly, Arbuscular mycorrhizal fungus (AMF) increases the absorption of phosphorus and other stable nutrients, improves plant tolerance in drought and soil salinity. These biocontrol agents provide an effective alternative to chemical pesticides, reduce the risk of insect resistance development, and support integrated pest management (IPM) strategy. The use of microbial inoculants has also accelerated the organic and regenerative farming system, where inputs should be natural, durable and residual free.

Plant-Microbe Interactions for Soil Health

Rhizosphere (a clay zone around the soil)- is a dynamic interface where plants and microbes are included in complex biochemical signaling, nutrient exchange and immune reactions. Beneficial microbes in rhizosphere contribute to the formation of overall, mineral weather and decomposition of organic matter, all needed for the formation and fertility of the soil. Some microorganisms increase the development and nutrients of the roots(Singh et al., 2024).

Others produce siderophores and antibiotics that suppress pathogenic microorganisms and protect the plants from diseases. These synergistic interactions form growth-promoting plants of rhizobacteria (PGPR) and endophytic microorganisms, which live in plant tissue without damaging. In recent years, progress in metagenomics and microbial ecology has enabled to better understand the microbiomes of clay - a diverse community of microorganisms in the soil - and its influence on plant health and ecosystem operations.

Studies have revealed that high microbial variations in the soil and functional redundancy system contribute to elasticity, nutrient cycling and carbon septation. Microbiome engineering, which includes manipulation in the microbial population to favor beneficial species or functions, is now being discovered as a border approach to precision agriculture. By customizing microbial consortia corresponding to specific crops and atmosphere.

Bioremediation and Sustainable Soil Management

In addition to the growth of crop growth and the availability of nutrients, soil biotechnology contributes significantly to bioremediation and durable soil management. Bioremediation is the use of organisms - mainly to detoxify and restore microorganisms and fungi - contaminated or degenerated soil. Through metabolic processes, these organisms can degrade organic pollutants such as petroleum hydrocarbons, pesticides and industrial dysfunction, or stabilize heavy metals by transforming in biosorption, bioaccumulation or low toxic forms (Martínez et al., 2025).

In agricultural settings, bioremediation provides cost-effective and environmentally friendly solutions to pesticide overs, excessive fertilizer application and contamination from agriculture-industrial debris. Simultaneously, sustainable soil management methods supported by biotechnology include composting, vermiculture, biochar application and green fertilizer integration. These practices increase the content of organic matter, improve soil formation and water-captivating ability, and promote a rich soil microbial ecosystem. The integration of microbial-based technologies with defense agricultural techniques-covers agriculture, crop rotation and cover-it expands their benefits, promoting land regeneration and climate elasticity (Pandiyan et al., 2024).

In conclusion, Microbial and clay biotechnology presents a powerful set of tools to promote durable, elastic and low-in-in-in-in-in-in-agricultural systems. By the use of biofertilizers, biopesticides, and beneficial plant-microbe interactions, farmers can reduce dependence on artificial inputs, increase crop influence, and preserve important soil ecosystems. Moreover, bioremediation and microbiome management provide strategic solutions to establish soil health in a degraded or polluted environment. As global focus changes toward regenerated agricultural and climate-smart practices, microbial and clay biotechnology will play an increasingly central role in shaping the future of agronomy. Its potential, microbial research, farmer training and scalable, to fully use the ongoing investment in the development of locally favorably bio-inputs. Ultimately, integrating these biological -based innovations into the mainstream agriculture not only supports productivity, but also adjusts with widespread goals of environmental stability and food system stability.

Figure 5. Digital integration and precision agronomy

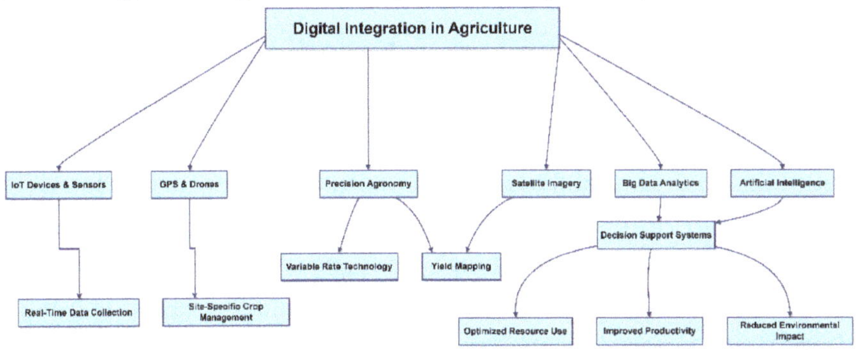

DIGITAL INTEGRATION AND PRECISION AGRONOMY

Bioinformatics and Genomic Data Analytics

Conversion of digital techniques with biotechnology reshapes modern agronomy, changes to data-jaws, precision-oriented and highly efficient farming systems. Digital integration in agronomy contains optimizing, genomic data analytics, artificial intelligence (AI), machine learning and application of IOT-enabled smart agricultural tools. The most transformative development plays the role of bioinformatics and genomic data analytics, which enables researchers to handle a large amount of genetic information produced by the high-throughput sequencing platform. Figure 5 illustrates digital integration and precision agronomy(Ali et al., 2024; Zhang et al., 2025).

Bioinformatics tools allow for comparisons, OT notation and comparison to full genome, important agriculture such as yields, stress tolerance and immunity. This simplifies the identity of genes associated with symptoms. Genome-Wide Association Studies (GWS), Quantitative Feature Locus (QTL) Mapping and Transcriptomic Profiling are now supported in breading pipelines, supported by cloud computing and advanced statistical models. These data -based approaches enable breeders to find genetic variations in different population, find novel alleles, and more accurately understand the gene function and regulatory network.

AI-Driven Decision Support in Breeding Programs

In view of this foundation, AI-powered decision support systems are revolutionizing how breeding programs operate and optimize. The machine can analyze multi-dimensional datasets, including genotype, phenotype, environmental and management data, to identify the patterns and predict breeding results with more accuracy than traditional methods. These models simplify the priority of the cross, estimation of breeding values, and the forecast of genotype exhibition under various environmental views. Deep learning tools are also employed to automate phenotyping using high-resolution images captured by drones, satellites or ground-based sensors(Kang, 2024; Zhang et al., 2025).

This digital phenotyping feature increases the accuracy of measurement and reduces the time and labor required for field tests. AI -powered platforms can integrate these data with climate models and clay maps so that the best genotypes are recommended for a specific agroecological zone, where there is an increase in the efficiency of varietal deployment. For example, the International Corn and Wheat Improvement Center (CIMMYT) helps platform breeders such as the genomic selection and decision support tool that balances potential, stress elasticity and market preferences. In essence, AI acts as a catalyst, transforms complex breeding data into functional insights that improve decision-making, reduce time-to-market, and increase genetic gains per unit time and cost.

Smart Agriculture and Biotech Synergy

The synergy between Smart Agriculture and Biotechnology is creating a new example in agronomic management, where real-time data and biological insights are informed about decisions. Smart agriculture operates a network of sensors, GPS technologies, drones and automated systems to monitor the development of crops, health of the soil, weather conditions and insect dynamics. When genetically modified seeds, microbial inoculants and biofortified crops are integrated with biotechnological inputs, these tools ensure that the interventions are definitely applied(Chandra et al., 2024).

Smart irrigation systems aware by clay sensors can pt over the use of water in drought-moisturizing GM crops, while drone-supported crops can detect the outbreak of early disease, enabling timely application of biopesticides. Precision nutrient management systems can correspond to fertilizer application based on real-time plants and earthen data, increase the functioning of biopesticides and reduce input waste. Moreover, blockchain and trace book techniques are being used to create trust and transparency in the agricultural value chains, the integrity and tracking of biotech seed products. This promotes stability by reducing the digital-biotech synergy input overuses, reducing environmental effects and maximizing resource-use efficiency.

Hence, Digital integration and precision agronomy is redefining the boundaries of what is possible in the novelty. By bioinformatics, AI and smart technologies, data -based decisions are becoming standard in breeding and farm management. These tools complement and expand the benefits of biotechnology, making agricultural systems more intelligent, adaptive and durable. As the access to digital infrastructure increases in stakeholders and data literacy increases, the possibility of this integration to transform global agriculture and support the climate-location will become more and more clear.

SOCIO-ECONOMIC AND ENVIRONMENTAL IMPACTS

The adoption of genetic engineering and biotechnology in Agronomy has created significant socio-economic and environmental influences both positive and fighting. From a socio-economic point of view, these technologies have contributed to increased agricultural productivity, especially in developing regions where biotechnological innovations such as BT cotton and drought-relevant maize have improved crop yields, loss of crop loss and farmer's income. Reducing dependence on artificial pesticides and fertilizers, input costs and advanced profitability in biotech crops have decreased, especially for small holder and resource-limited farmers(Olesen, 2016).

In addition, the introduction of nutritious crops, such as gold rice or iron-fortified lumps, supports public health objectives by removing subtle nutrients in the sensitive population. On the employment front, biotechnology-based agriculture has stimulated the formation of jobs in seed production, Agri-input distribution and bio-based production development. However, in the margins of technical accesses, corporate control of seeds and traditional knowledge system, there is concern over raising issues around equity and farmer autonomy(Fischer et al., 2005; Mannion & Morse, 2013).

Environmentally, biotechnology has contributed to a more sustainable farming system by reducing chemical consumption, protecting water by drought-over-drought varieties, and enabling no-tall practices that maintain soil composition and carbon content. Genetically improved crops have played a role in integrated pest management (IPM), reducing pesticide runs and enhancing agroecological balance. Nevertheless, environmental risks remain under scrutiny, including the possibility of gene flow for wild relatives, development of pesticides and unwanted effects on non-target organisms. Long -term ecological assessments and solid biopsy rules are required to reduce these concerns. Overall, when biotechnology offers significant socio-economic and environmental benefits, transparent regime for its proper deployment, farmer-centered innovation models delays and policy framework are needed to ensure that its benefits are divided into all sections of the society and sets.

REGULATORY, ETHICAL, AND PUBLIC PERCEPTION ISSUES

The implementation of genetic engineering and biotechnology in Agronomy is closely operated by a complex web of regulatory, moral and public perceptions. In some countries, adopting regulatory, adoption-based approaches, when others apply restricted policies based on the principles of caution, vary significantly. Agencies like USDA, EFSA and India's GEAC monitors the approval and monitoring of genetically modified organisms (GMOs) focusing on biopsy, environmental influence and food security. Diversion in regulatory standards often creates trade barriers, delays innovation, and complicate technical deployment. The arrival of genome editing technologies such as CRISPR-CAS has further challenged the existing framework, as these technologies often do not include transgenic changes and traditional GMOs may be out of definitions, which reflects the variable progress(Brookes et al., 2006).

Morally, biotechnology concerns about owning genetic resources, patenting life forms and potential erosion of traditional farming methods. In some multinational corporations, the concentration of intellectual property rights has discussed the right distribution of seed sovereignty, farmer rights and technical benefits. There are also moral questions regarding the change of natural genome, cross-species gene transfer and long-term ecological influences that are still fully understood. These concerns require dialogues included in the containing moralists, farmers, and scientists.

Public assumptions play a crucial role in adopting agricultural biotechnology. Despite the unanimous consensus on the safety of many biotech crops in false information, cultural beliefs and mistrust in regulatory institutions. Effective Vijay Communication requires communication, transparent risk assessments and engagement of stakeholders to create public confidence. Ultimately, addressing the regulatory, moral and vision issues with clarity, ness painting and incorporation is the key to the responsible and fair progress of agricultural biotechnology.

FUTURE DIRECTIONS AND STRATEGIC OUTLOOK

Trends Shaping the Next Decade

As weather change in agriculture, lack of resources, population growth and changing diet, genetic engineering and biotechnology in agronomy faces increasing pressure on their ability to operate, elastic, efficient and incorporated food systems. Over the next decade, many transformative trends are expected to shape this area. Genome editing technologies (especially CRISPRs and its derivatives), large data, artificial intelligence and artificial life will increase the precision, scalability and adaptability of crop correction applications. Progress in multi- omics integration (genomics, proteomics, transcriptomics, and metabolomics) will provide ER plague insights in plant biology, enable feature stacking and customized crop design. Genomic selection and expansion of digital breeding platforms will allow real-time decisions and adaptive breeding corresponding to the local environment. Moreover, climate-smart biotech will be the center for achieving agricultural stability in engineer-extreme conditions for crop-carbon efficiency, water productivity and heat elasticity(Zhang et al., 2025).

Policy Recommendations and Innovation Roadmaps

In parallel, Effective policy recommendations and innovation route maps will be important to ensure responsible deployment and broad-based benefits. Policy partners must prioritize the expressions. The based, transparent and harmonious regulatory structures that balance innovation with biopsy. GMOs can promote fast, safe innovation to include precision genome-traded products, streamline approval processes, and adopt tier risk assessment models to update definitions. Biotechnology should be strengthened to make public sector investment, especially in developing countries, to create local capacity and reduce dependence on multinational companies. The structure of the intellectual property must also support proper entry, especially for small holder farmers and public breeding organizations. Promote a public-private partnership, open-source biotech platform and participation breeding Models Delo to make innovation democracy. Moreover, embedding moral considerations, gender incorporation and indigenous knowledge in R&D planning will ensure that biotech innovations align with socio-cultural values and human rights(Stewart Jr, 2025a, 2025b).

Toward Resilient and Inclusive Agronomic Systems

Looking forward, creating elastic and incorporated agronomic systems requires an integrated approach that moves beyond technical innovation.

Biotechnology must be combined to regenerate the soil with agrodolces principles, regenerated farming methods, and circular bioeconomy models, to protect biodiversity and reduce greenhouse gas emissions. The adoption of biotech-uplift crops should match the infrastructure, farmer training and digital advisory services to ensure successful integration in the farming system(Ali et al., 2025; Martínez et al., 2025).

The incorporation biotech research, decisions and benefits will also depend on the empowerment of women, youth and margins. Global cooperation in governments, research institutions, development agencies and civil society will be required to collect knowledge, money and regime for similar results. In conclusion, the future of biotechnology in agronomy is to create a strong ecosystem, not only in the most enemy. In the effect of innovation, but with the effect of innovation. By accepting data -based breeding, regulatory improvement, moral regime and inclusion of bottom, biotechnology can be part of an elastic, fair and durable global food system.

CONCLUSION

Genetic engineering and biotechnology modern agriculture. The transforming into the transforming of some of the most challenges of agriculture has become important in providing innovative, precise and sustainable solutions. Biotechnology has expanded the boundaries of traditional plant breeding, to enhancing the elasticity of the crop, to environmental stress and controlling pests and diseases, enriching the nutritional profile of the food crop.

The integration of digital tools expands biotechnology influence by enabling Bioinformatics, AI-powered analytics, and smart agriculture platform-data-unifier decision-determination, real-time monitoring and corresponding agronomic intervention. This harmonious precision between organisms and technology is redefining the Agronomy and setting the stage for climate-smart, resource-efficient and adaptive farming system.

However, the success and acceptance of these technologies is the same not only on Vijay's progress but also on responsible regime, equal access and moral deployment. Regulatory clarity, transparent risk communication, incorporated innovation models delts and public engagements are required to ensure that biotechnology delivers its promise without enhancing social or environmental inequality. Considering the concerns around beasty, corporate control and ecological effects, it would be key to making public confidence and long -term stability.

REFERENCES

Abdul Aziz, M., Brini, F., Rouached, H., & Masmoudi, K. (2022). Genetically engineered crops for sustainably enhanced food production systems. *Frontiers in Plant Science, 13*, 1027828. DOI: 10.3389/fpls.2022.1027828 PMID: 36426158

Adhikari, D. (2024). Plant Heat Shock Proteins-Origin, Evolution and Mechanism of Action for Biotechnological Implications in Crop Improvement. [IJIRSET]. *International Journal of Innovative Research in Science, Engineering and Technology, 13*(3).

Ali, A., Altaf, M. T., Bedir, M., Liaqat, W., Jamil, A., Nadeem, M. A., & Baloch, F. S. (2024). Biotechnological advancements toward sugarcane crop improvement. In *Industrial Crop Plants* (pp. 333–352). Springer. DOI: 10.1007/978-981-97-1003-4_13

Ali, A., Jabeen, N., Farruhbek, R., Chachar, Z., Laghari, A. A., Chachar, S., Ahmed, N., Ahmed, S., & Yang, Z. (2025). Enhancing nitrogen use efficiency in agriculture by integrating agronomic practices and genetic advances. *Frontiers in Plant Science, 16*, 1543714. DOI: 10.3389/fpls.2025.1543714 PMID: 40161228

Bharat, R. A., Prathmesh, S. P., Sarsu, F., & Suprasanna, P. (2024). Induced mutagenesis using gamma rays: Biological features and applications in crop improvement. *OBM Genetics, 8*(2), 1–27. DOI: 10.21926/obm.genet.2402233

Brookes, G., & Barfoot, P. (2006). Global impact of biotech crops: Socio-economic and environmental effects in the first ten years of commercial use. *AgBioForum, 9*(3), 139–151.

Chandra, K., Chand, S., Saini, R. P., & Sharma, R. (2024). *Smart breeding: Molecular interventions and advancements for crop improvement*. CRC Press. DOI: 10.1201/9781003361862

Demirer, G. S., Silva, T. N., Jackson, C. T., Thomas, J. B. W., Ehrhardt, D., Rhee, S. Y., Mortimer, J. C., & Landry, M. P. (2021a). Nanotechnology to advance CRISPR–Cas genetic engineering of plants. *Nature Nanotechnology*, *16*(3), 243–250. DOI: 10.1038/s41565-021-00854-y PMID: 33712738

Demirer, G. S., Silva, T. N., Jackson, C. T., Thomas, J. B. W., Ehrhardt, D., Rhee, S. Y., Mortimer, J. C., & Landry, M. P. (2021b). Nanotechnology to advance CRISPR–Cas genetic engineering of plants. *Nature Nanotechnology*, *16*(3), 243–250. DOI: 10.1038/s41565-021-00854-y PMID: 33712738

Fischer, G., Shah, M. N., Tubiello, F., & Van Velhuizen, H. (2005). Socio-economic and climate change impacts on agriculture: An integrated assessment, 1990–2080. *Philosophical Transactions of the Royal Society of London. Series B, Biological Sciences*, *360*(1463), 2067–2083. DOI: 10.1098/rstb.2005.1744 PMID: 16433094

Gao, C. (2021). Genome engineering for crop improvement and future agriculture. *Cell*, *184*(6), 1621–1635. DOI: 10.1016/j.cell.2021.01.005 PMID: 33581057

Hamdan, M. F., Mohd Noor, S. N., Abd-Aziz, N., Pua, T.-L., & Tan, B. C. (2022). Green revolution to gene revolution: Technological advances in agriculture to feed the world. *Plants*, *11*(10), 1297. DOI: 10.3390/plants11101297 PMID: 35631721

Kalendar, R., Orbovic, V., Egea-Cortines, M., & Song, G. (2022). Recent advances in plant genetic engineering and innovative applications. *Frontiers in Plant Science*, *13*, 1045417. DOI: 10.3389/fpls.2022.1045417 PMID: 36340337

Kang, M. S. (2024). *Crop improvement: Challenges in the twenty-first century*. CRC Press. DOI: 10.1201/9781003578512

Lallawmkimi, M. C., Veda, D. S., Yadav, A., Kumar, M., & Rout, A. (2024). Innovative approaches in crop genetic engineering for sustainable agriculture: A review. *Journal of Advances in Biology & Biotechnology*, *27*(8), 615–631. DOI: 10.9734/jabb/2024/v27i81177

Mamata, K., & Lamichhane, A. (2021). Advances in agricultural biotechnology. *Nepal Journal of Biotechnology*, *9*(1), 85–92. DOI: 10.3126/njb.v9i1.38643

Mannion, A., & Morse, S. (2013). GM crops 1996–2012: A review of agronomic, environmental and socio-economic impacts. *University of Surrey, Centre for Environmental Strategy Working Paper, 4*(13), 1–40.

Martínez, R. T., Cruz-Martinez, Y. R., Villagómez-Guzmán, A. K., Pérez-Valera, O., Arreaga-González, H. M., & Hernández-Delgado, T. (2025). Biofertilizers and Biopesticides: Sustainable Alternatives for Agriculture. In *Soil Improvement and Water Conservation Biotechnology* (pp. 113–136). Bentham Science Publishers.

Olesen, J. E. (2016). Socio-economic impacts—Agricultural systems. *North Sea Region Climate Change Assessment*, 397–407.

Pandiyan, A., Sarsan, S., Durga, G. G. S., & Ravikumar, H. (2024). Biofertilizers and biopesticides as microbial inoculants in integrated pest management for sustainable agriculture. In *Microbial essentialism* (pp. 485–518). Elsevier. DOI: 10.1016/B978-0-443-13932-1.00010-6

Singh, S., Ashoka, P., Ahlawat, U., Changdeo, W. B., Rehsawla, R., Naruka, A., & Sharma, D. (2024). Mechanisms and Applications of Microbial Biotechnology in Soil Health and Agricultural Productivity: A Review. *Journal of Advances in Biology & Biotechnology*, *27*(7), 1420–1438. DOI: 10.9734/jabb/2024/v27i71104

Slafer, G. A., Molina-Cano, J. L., Savin, R., Araus, J. L., & Romagosa, I. (2024). *Barley science: Recent advances from molecular biology to agronomy of yield and quality*.

Squire, H. J., Tomatz, S., Voke, E., González-Grandío, E., & Landry, M. (2023a). The emerging role of nanotechnology in plant genetic engineering. *Nature Reviews Bioengineering*, *1*(5), 314–328. DOI: 10.1038/s44222-023-00037-5

Squire, H. J., Tomatz, S., Voke, E., González-Grandío, E., & Landry, M. (2023b). The emerging role of nanotechnology in plant genetic engineering. *Nature Reviews Bioengineering*, *1*(5), 314–328. DOI: 10.1038/s44222-023-00037-5

Stewart, C. N.Jr. (2025a). *Plant biotechnology and genetics: Principles, techniques, and applications*. John Wiley & Sons.

Stewart, C. N.Jr. (2025b). *Plant biotechnology and genetics: Principles, techniques, and applications*. John Wiley & Sons.

Tyczewska, A., Twardowski, T., & Woźniak-Gientka, E. (2023). Agricultural biotechnology for sustainable food security. *Trends in Biotechnology*, *41*(3), 331–341. DOI: 10.1016/j.tibtech.2022.12.013 PMID: 36710131

Upadhyay, S. K. (2021). *Genome engineering for crop improvement*. Wiley Online Library. DOI: 10.1002/9781119672425

Verma, K. K., Song, X.-P., Budeguer, F., Nikpay, A., Enrique, R., Singh, M., Zhang, B.-Q., Wu, J.-M., & Li, Y.-R. (2022). Genetic engineering: An efficient approach to mitigating biotic and abiotic stresses in sugarcane cultivation. *Plant Signaling & Behavior*, *17*(1), 2108253. DOI: 10.1080/15592324.2022.2108253 PMID: 35959678

Yan, Y., Zhu, X., Yu, Y., Li, C., Zhang, Z., & Wang, F. (2022). Nanotechnology strategies for plant genetic engineering. *Advanced Materials*, *34*(7), 2106945. DOI: 10.1002/adma.202106945 PMID: 34699644

Zhang, D., Xu, F., Wang, F., Le, L., & Pu, L. (2025). Synthetic biology and artificial intelligence in crop improvement. *Plant Communications*, *6*(2), 101220. DOI: 10.1016/j.xplc.2024.101220 PMID: 39668563

Compilation of References

Abdel-Fattah, M. K. (2019). Reclamation of saline-sodic soils for sustainable agriculture in Egypt. Sustainability of Agricultural Environment in Egypt: Part II: Soil-Water-Plant Nexus, 69-92.

Abdul Aziz, M., Brini, F., Rouached, H., & Masmoudi, K. (2022). Genetically engineered crops for sustainably enhanced food production systems. *Frontiers in Plant Science*, *13*, 1027828. DOI: 10.3389/fpls.2022.1027828 PMID: 36426158

Adhikari, D. (2024). Plant Heat Shock Proteins-Origin, Evolution and Mechanism of Action for Biotechnological Implications in Crop Improvement. [IJIRSET]. *International Journal of Innovative Research in Science, Engineering and Technology*, *13*(3).

Afek, U., Orenstein, J., & Kim, J. J. (2001). Control of silver scurf disease in stored potato by using hydrogen peroxide plus (HPP). *Crop Protection (Guildford, Surrey)*, *20*(1), 69–71. DOI: 10.1016/S0261-2194(00)00054-5

Afzal, I., & Javed, T., Amirkhani. (2020). Modern technologies of the sowing system: Application of coatings on seeds to increase seed yield and yield. *Agriculture*, *10*, 526. DOI: 10.3390/agriculture10110526

Agegnehu, G., & Amede, T. (2017). Integrated soil fertility and plant nutrient management in tropical agro-ecosystems: A review. *Pedosphere*, *27*(4), 662–680. DOI: 10.1016/S1002-0160(17)60382-5

Aggarwal, P. K. (2008). Global climate change and Indian agriculture: Impacts, adaptation and mitigation. *Indian Journal of Agricultural Sciences*, *78*(11), 911.

Aggarwal, P. K., Joshi, H. C., Singh, S. D., Bhatia, A., Jain, N., Prasad, S., Chaudhary, A., Gupta, N., & Pathak, H. (2009). Agriculture and environment. In *Hand Book of Agriculture, Directorate of Information and Publication Agriculture* (pp. 62–92). ICAR.

Aggarwal, P. K., Singh, A. K., Samra, J. S., Singh, G., Gogoi, A. K., Rao, G. G. S. N., & Ramakrishna, Y. S. (2009). Introduction. In Aggarwal, P. K. (Ed.), *Global Climate Change and Indian Agriculture* (pp. 1–5). ICAR.

Akhigbe, B., Munir, K., Akinade, O., Akanbi, L., & Oyedele, L. (2021). IoT Technologies for Livestock Management: A Review of Present Status, Opportunities, and Future Trends. *Big Data and Cognitive Computing*, *5*(1), 10. DOI: 10.3390/bdcc5010010

Alexander, L. V., Zhang, X., Peterson, T. C., Caesar, J., Gleason, B., Klein Tank, A. M. G., Haylock, M., Collins, D., Trewin, B., Rahimzadeh, F., Tagipour, A., Rupa Kumar, K., Revadekar, J., Griffiths, G., Vincent, L., Stephenson, D. B., Burn, J., Aguilar, E., Brunet, M., & Vazquez-Aguirre, J. L. (2006). Global observed changes in daily climate extremes of temperature and precipitation. *Journal of Geophysical Research*, *111*(D5), 2005JD006290. DOI: 10.1029/2005JD006290

Ali, A., Altaf, M. T., Bedir, M., Liaqat, W., Jamil, A., Nadeem, M. A., & Baloch, F. S. (2024). Biotechnological advancements toward sugarcane crop improvement. In *Industrial Crop Plants* (pp. 333–352). Springer. DOI: 10.1007/978-981-97-1003-4_13

Ali, A., Jabeen, N., Farruhbek, R., Chachar, Z., Laghari, A. A., Chachar, S., Ahmed, N., Ahmed, S., & Yang, Z. (2025). Enhancing nitrogen use efficiency in agriculture by integrating agronomic practices and genetic advances. *Frontiers in Plant Science*, *16*, 1543714. DOI: 10.3389/fpls.2025.1543714 PMID: 40161228

Anagnostis, A., Papageorgiou, E., & Bochtis, D. (2020). *Application* of Artificial Neural Networks for Natural Gas Consumption Forecasting. *Sustainability (Basel), 12*(16), 6409. DOI: 10.3390/su12166409

Anagnostis, A., Tagarakis, A. C., Asiminari, G., Papageorgiou, E., Kateris, D., Moshou, D., & Bochtis, D. (2021). A deep learning approach for anthracnose infected trees classification in walnut orchards. *Computers and Electronics in Agriculture, 182*, 105998. DOI: 10.1016/j.compag.2021.105998

Ando, K., & Grumet, R. (2006). Evaluation of altered cucumber plant architecture as a means to reduce *Phytophthora capsici* disease incidence on cucumber fruit. *Journal of the American Society for Horticultural Science, 131*(4), 491–498. DOI: 10.21273/JASHS.131.4.491

Annamalai, H. (2010). Moist dynamical linkage between the equatorial Indian Ocean and the South Asian monsoon trough. *Journal of the Atmospheric Sciences, 67*(3), 589–610. DOI: 10.1175/2009JAS2991.1

Anonymous. (2019). *Potential of ICT in water sector*. https://www.teriin.org/events/icdl/sdg-files/pdf/ICT-in-water-sector.pdf

Anwar, A., Xianchang, Yu., & Jansu, L. I. (2020). Soaking seeds as a promising method to improve growth, chlorophyll, photosynthesis and nutrient content in cucumber seedlings. *Notulae Botanicae Horti Agrobotanici Cluj-Napoca, 48*(1), 116–127. DOI: 10.15835/nbha48111806

Ayers, R. S., & Westcot, D. W. (1985). Water quality for agriculture (Vol. 29, p. 174). Rome: Food and agriculture organization of the United Nations.

Aylmore, L. A. G., Karim, M., & Quirk, J. P. (1971). Dissolution of gypsum, monocalcium phosphate and superphosphate fertilizers in relation to particle size and porous structure. *Soil Research (Collingwood, Vic.), 9*(1), 21–32.

Bahl, P. N. (2015). Climate change and pulses: Approaches to combat its impact. *Agricultural Research, 4*(2), 103–108. DOI: 10.1007/s40003-015-0163-9

Balla, A., Silini, A., Cherif-Silini, H., Chenari Bouket, A., Alenezi, F. N., & Belbahri, L. (2022). Recent advances in encapsulation techniques of plant growth-promoting microorganisms and their prospects in the sustainable agriculture. *Applied Sciences (Basel, Switzerland)*, *12*(18), 9020. DOI: 10.3390/app12189020

Bandyopadhyay, B. K., Maji, B., Sen, H. S., & Tyagi, N. K. (2003). Coastal soils of West Bengal—Their nature, distribution and characteristics. *Bulletin*, *1*, 1–62.

Bañón, S., Ochoa, J., Bañón, D., Ortuño, M. F., & Sánchez-Blanco, M. J. (2019). Controlling salt flushing using a salinity index obtained by soil dielectric sensors improves the physiological status and quality of potted hydrangea plant. *Scientia Horticulturae*, *247*, 335–343. DOI: 10.1016/j.scienta.2018.12.026

Barkunan, S. R., Bhanumathi, V., & Sethuram, J. (2019). Smart sensor for automatic drip irrigation system for paddy cultivation. *Computers & Electrical Engineering*, *73*, 180–193. DOI: 10.1016/j.compeleceng.2018.11.013

Benazzuk, S., Jazuli, Z.E., Lutts, S. (2019). Vermicompost filtrate as a promising means for priming and rejuvenation of germinating Brassica napus seeds treated with salt. *Communes. Soil science. Vegetable anal.*, 50, 1344–1357.

Benos, L., Bechar, A., & Bochtis, D. (2020). Safety, and ergonomics in human-robot interactive agricultural operations. *Biosystems Engineering*, *200*, 55–72. DOI: 10.1016/j.biosystemseng.2020.09.009

Benouniche, M., Kuper, M., Hammani, A., & Boesveld, H. (2014). Making the user visible: Analysing irrigation practices and farmers' logic to explain actual drip irrigation performance. *Irrigation Science*, *32*(6), 405–420. DOI: 10.1007/s00271-014-0438-0

Bernstein, L., & Fireman, M. (1957). Laboratoty Studies on Salt Distribution in Furrow-Irrigated Soil With Special Reference to the Pre-Emergence Period. *Soil Science*, *83*(4), 249–264. DOI: 10.1097/00010694-195704000-00001

Bharat, R. A., Prathmesh, S. P., Sarsu, F., & Suprasanna, P. (2024). Induced mutagenesis using gamma rays: Biological features and applications in crop improvement. *OBM Genetics*, *8*(2), 1 27. DOI: 10.21926/obm.genet.2402233

Bimpizas-Pinis, M., Santagata, R., Kaiser, S., Liu, Y., & Lyu, Y. (2022). Additives in the food supply chain: Environmental assessment and circular economy implications. *Environmental and Sustainability Indicators*, *14*, 100172. DOI: 10.1016/j.indic.2022.100172

Blandford, D. and Josling, T. (2009). Greenhouse gas reduction policies and agriculture: implications for production incentives and international trade disciplines.

Boddy, L. (2016). Fungi, ecosystems and global change. In *The fungi* (pp. 361–400). Academic Press. DOI: 10.1016/B978-0-12-382034-1.00011-6

Bonnet, P., Joly, A., Goëau, H., Champ, J., Vignau, C., Molino, J. F., Barthélémy, D., & Boujemaa, N. (2016). Plant identification: Man vs. machine. *Multimedia Tools and Applications*, *75*(3), 1647–1665. DOI: 10.1007/s11042-015-2607-4

Bradford, KJ, Steiner, JJ, Travata, S.E. (1990). Analysis of seed seedling for germination and appearance of batches of pepper seeds. *The science of crop production,* 30, 718–721.

Brookes, G., & Barfoot, P. (2006). Global impact of biotech crops: Socio-economic and environmental effects in the first ten years of commercial use. *AgBioForum*, *9*(3), 139–151.

Bucka, F., Kölbl, A., Marschner, P., Fitzpatrick, R., Mosley, L., & Kögel-Knabner, I. (2018). Remediation of acid sulphate soils by addition of organic matter. In EGU General Assembly Conference Abstracts (p. 2378).

Burman, D., Bandyopadhyay, B. K., & Mahanta, K. K. (2010). Management of acid sulphate soil of coastal Sundarbans region: Observations under on-farm trial. *Journal of the Indian Society of Coastal Agricultural Research*, *28*(1), 8–11.

Cai, W., Cowan, T., & Thatcher, M. (2012). Rainfall reductions over Southern Hemisphere semi-arid regions: The role of subtropical dry zone expansion. *Scientific Reports*, *2*(1), 702. DOI: 10.1038/srep00702

Carter, D. L., & Fanning, C. D. (1964). Combining surface mulches and periodic water applications for reclaiming saline soils. *Soil Science Society of America Journal*, *28*(4), 564–567. DOI: 10.2136/sssaj1964.03615995002800040033x

Chandra, K., Chand, S., Saini, R. P., & Sharma, R. (2024). *Smart breeding: Molecular interventions and advancements for crop improvement.* CRC Press. DOI: 10.1201/9781003361862

Chawla, K. L., & Abrol, I. P. (1982). Effect of gypsum fineness on the reclamation of sodic soils. *Agricultural Water Management*, *5*(1), 41–50. DOI: 10.1016/0378-3774(82)90037-3

Chhachhar, A. R., Qureshi, B., Khushk, G. M., & Maher, Z. A. (2014). Use of mobile phone among farmers for agriculture information. *European Journal of Scientific Research*, *119*(2), 265–271.

Cho, K. M., Ranamukhaarachchi, S. L., & Zoebisch, M. A. (2002). Cropping systems on acid sulphate soils in the central plains of Thailand: constraints and remedies. 17th WCSS, 14-21.

Choudhary, A. K., Amrate, P. K., & Chatterjee, A. (2014). An unusual incidence and population dynamics of sugarcane leaf hopper, *Pyrilla perpusilla* (Walk) and its natural enemies in Madhya Pradesh, India. *Environmental Conservation*, *20*, 337–342.

Choudhary, V. K., & Kumar, P. S. (2013). Crop and water productivity, profitability and energy consumption pattern of a maize-based crop sequence in the North Eastern Himalayan Region, India. *Archives of Agronomy and Soil Science*, *59*(5), 653–669. DOI: 10.1080/03650340.2012.665992

Chowdhury, H. (2023). Circular economy integration in additive manufacturing. *Preprints, Oct, 3*.

Cicculло, F., Cagliano, R., Bartezzaghi, G., & Perego, A. (2021). Implementing the circular economy paradigm in the agri-food supply chain: The role of food waste prevention technologies. *Resources, Conservation and Recycling*, *164*, 105114. DOI: 10.1016/j.resconrec.2020.105114

Conrad, Z., Niles, M. T., Neher, D. A., Roy, E. D., Tichenor, N. E., & Jahns, L. (2018). Relationship between food waste, diet quality, and environmental sustainability. *PLoS One*, *13*(4), e0195405. DOI: 10.1371/journal.pone.0195405 PMID: 29668732

Dahikar, M. S. S., & Rode, S. V. (2014). Agricultural crop yield prediction using an artificial neural network approach. *International Journal of Innovative Research in Electrical, Electronics, Instrumentation, and Control Engineering*, *2*(1), 1–4.

Dalhem, K., Engblom, S., Stén, P., & Österholm, P. (2019). Subsurface hydrochemical precision treatment of a coastal acid sulfate soil. *Applied Geochemistry*, *100*, 352–362. DOI: 10.1016/j.apgeochem.2018.12.005

Damalas, C., Koutroubas, S. D., & Fotiadis, S. (2019). The effect of the hydroprimer on seed germination and field productivity of horse beans during spring sowing. *Agriculture*, *9*(9), 201. DOI: 10.3390/agriculture9090201

Das, S. K., & Das, S. K. (2015). Acid sulphate soil: management strategy for soil health and productivity. Popular kheti, 3(2), 2-7.

Dasgupta, M. K. (1988). *Principles of plant pathology*. Allied Publishers.

Das, S., Munshi, M. N., & Kabir, W. (2017). The impact of ICTs on agricultural production in Bangladesh: A study with food crops. *SAARC Journal of Agriculture*, *14*(2), 78–89. DOI: 10.3329/sja.v14i2.31247

Demirer, G. S., Silva, T. N., Jackson, C. T., Thomas, J. B. W., Ehrhardt, D., Rhee, S. Y., Mortimer, J. C., & Landry, M. P. (2021a). Nanotechnology to advance CRISPR–Cas genetic engineering of plants. *Nature Nanotechnology*, *16*(3), 243–250. DOI: 10.1038/s41565-021-00854-y PMID: 33712738

Demmers, T. G., Cao, Y., Parsons, D. J., Gauss, S., & Wathes, C. M. (2012). Simultaneous Monitoring and Control of Pig Growth and Ammonia Emissions. *IX International Livestock Environment Symposium (ILES IX)*. Valencia, Spain: American Society of Agricultural and Biological Engineers. DOI: 10.13031/2013.41560

Dent, D. L. (1986). *Acid sulphate soils: a baseline for research and development (No. 39)*. ILRI.

Devika O.S., Singh S., Sarkar D., Barnwal P., Suman J. and Rakshit A. (2021). Seed seeding: a potential complement to integrated resource management in fragile intensive ecosystems. *The boundaries of sustainable food systems*, 5. .DOI: 10.3389/fsufs.2021.654001

Dey, S., Srivastava, P. K., Maji, S., Das, M. K., Mukhopadhyaya, M. K., & Saha, P. K. (2007). Impact of climate change on the breeding of Indian major carps in West Bengal. *Journal of the Inland Fisheries Society of India*, *39*(1), 26–34.

Doorn, N. (2021). Artificial intelligence in the water domain: Opportunities for responsible use. *The Science of the Total Environment*, *755*, 1–10. DOI: 10.1016/j.scitotenv.2020.142561 PMID: 33039891

Ebimol, N. L., Suresh, P. R., Binitha, N. K., & Santhi, G. R. (2017). Management of iron and aluminium toxicity in acid sulphate soils of Kuttanad. *International Journal of Current Microbiology and Applied Sciences*, *6*(11), 1496–1503. DOI: 10.20546/ijcmas.2017.611.177

El Bilali, A., Taleb, A., & Brouziyne, Y. (2021). Ground water quality forecasting using machine learning algorithms for irrigation purposes. *Agricultural Water Management*, *245*, 106625. DOI: 10.1016/j.agwat.2020.106625

El-Gabaly, M. M. (1971). Reclamation and management of salt affected soils. In *Salinity Seminar, Baghdad. Irrigation and Drainage Paper, 7*. FAO.

Fanning, D. S. (1993). Salinity problems in acid sulfate coastal soils. Towards the rational use of high salinity tolerant plants: Vol. 1 Deliberations about High Salinity Tolerant Plants and Ecosystems, 491-500. DOI: 10.1007/978-94-011-1858-3_52

Fischer, G., Shah, M. N., Tubiello, F., & Van Velhuizen, H. (2005). Socio-economic and climate change impacts on agriculture: An integrated assessment, 1990–2080. *Philosophical Transactions of the Royal Society of London. Series B, Biological Sciences*, *360*(1463), 2067–2083. DOI: 10.1098/rstb.2005.1744 PMID: 16433094

Food and Agriculture Organization of the United Nations. (2019). *Food and Agriculture Organization of the United Nations*. https://www.fao.org/fao-stories/article/en/c/1185405

Fuhrer, J. (2003). Agroecosystem responses to combinations of elevated CO_2, ozone and global climate change. *Agriculture, Ecosystems & Environment*, *97*(1-3), 1–20. DOI: 10.1016/S0167-8809(03)00125-7

Fu, X., Zheng, Z., Sha, Z., Cao, H., Yuan, Q., Yu, H., & Li, Q. (2022). Biorefining waste into nanobiotechnologies can revolutionize sustainable agriculture. *Trends in Biotechnology*, *40*(12), 1503–1518. DOI: 10.1016/j.tibtech.2022.09.013 PMID: 36270903

Gaba, Y., Pareek, A., & Singla-Pareek, S. L. (2021). Raising climate-resilient crops: Journey from the conventional breeding to new breeding approaches. *Current Genomics*, *22*(6), 450–467. DOI: 10.2174/1389202922666210928151247

Galvão, R. K. H., Araújo, M. C. U., Fragoso, W. D., Silva, E. C., José, G. E., Soares, S. F. C., & Paiva, H. M. (2008). A variable elimination method to improve the parsimony of MLR models using the successive projections algorithm. *Chemometrics and Intelligent Laboratory Systems*, *92*(1), 83–91. DOI: 10.1016/j.chemolab.2007.12.004

Gao, C. (2021). Genome engineering for crop improvement and future agriculture. *Cell*, *184*(6), 1621–1635. DOI: 10.1016/j.cell.2021.01.005 PMID: 33581057

Giakoumatos, S. D., & Tsoulfas, G. T. (2025). Towards Shaping Integrated Circular Economy Approaches in Animal Breeding Supply Chains by Mainstreaming Microalgae Utilization. *Circular Economy and Sustainability*, *5*(7), 1–41. DOI: 10.1007/s43615-025-00520-8

Gilligan, C. A., & van den Bosch, F. (2008). Epidemiological models for invasion and persistence of pathogens. *Annual Review of Phytopathology*, *46*(1), 385–418. DOI: 10.1146/annurev.phyto.45.062806.094357

Giri, G.S., Schillinger, V.F. (2003). Seedling of winter wheat for germination, germination and yield. *The science of crop production,* 43, 2135-2141.

Gora, M. K., Kumar, S., Jat, H. S., Kakraliya, S. K., Choudhary, M., Dhaka, A. K., Jat, R. D., Kakraliya, M., Sharma, P. C., & Jat, M. L. (2022). Scalable diversification options delivers sustainable and nutritious food in Indo-Gangetic plains. *Scientific Reports*, *12*(1), 14371. DOI: 10.1038/s41598-022-18156-1

Grieve, C. M., Grattan, S. R., & Maas, E. V. (2012). Plant salt tolerance. ASCE manual and reports on engineering practice, 71, 405-459.

Gu, Y. Y, James. W, McNicol. M, (1994). An Application of Belief Networks to Future Crop Production. *IEEE Conference on Artificial Intelligence for Applications, San Antonia, TX*. pp. 305.

Guggenberger, G. (2005). Humification and mineralization in soils. In *Microorganisms in soils: roles in genesis and functions* (pp. 85–106). Springer Berlin Heidelberg. DOI: 10.1007/3-540-26609-7_4

Gupta, A. K., Tyagi, P., & Sehgal, V. K. (2011). Drought disaster challenges and mitigation in India: Strategic appraisal. *Current Science*, •••, 1795–1806.

Gupta, A., & Pathak, H. (2016). *Climate change and agriculture in India.*

Hamdan, M. F., Mohd Noor, S. N., Abd-Aziz, N., Pua, T.-L., & Tan, B. C. (2022). Green revolution to gene revolution: Technological advances in agriculture to feed the world. *Plants*, *11*(10), 1297. DOI: 10.3390/plants11101297 PMID: 35631721

Harvey, C. A., Rakotobe, Z. L., Rao, N. S., Dave, R., Razafimahatratra, H., Rabarijohn, R. H., Rajaofara, H., & MacKinnon, J. L. (2014). Extreme vulnerability of smallholder farmers to agricultural risks and climate change in Madagascar. *Philosophical Transactions of the Royal Society of London. Series B, Biological Sciences*, *369*(1639), 369. DOI: 10.1098/rstb.2013.0089 PMID: 24535397

Hatfield, J. L., & Prueger, J. H. (2015). Temperature extremes: Effect on plant growth and development. *Weather and Climate Extremes*, *10*, 4–10. DOI: 10.1016/j.wace.2015.08.001

Helm, J. M., Swiergosz, A. M., Haeberle, H. S., Karnuta, J. M., Schaffer, J. L., Krebs, V. E., Spitzer, A. I., & Ramkumar, P. N. (2020). Machine Learning and Artificial Intelligence: Definitions, Applications, and Future Directions. *Current Reviews in Musculoskeletal Medicine*, *13*(1), 69–76. DOI: 10.1007/s12178-020-09600-8 PMID: 31983042

Henry, R. J. (2014). Genomics strategies for germplasm characterization and the development of climate resilient crops. *Frontiers in Plant Science*, *5*, 68. DOI: 10.3389/fpls.2014.00068

Hettiarachchi, B. D., Brandenburg, M., & Seuring, S. (2022). Connecting additive manufacturing to circular economy implementation strategies: Links, contingencies and causal loops. *International Journal of Production Economics*, *246*, 108414. DOI: 10.1016/j.ijpe.2022.108414

Hinojosa-Davalos, J., Arias-Rios, EV, & Varela-Hernandez, J.J., Cardona-Lopez, Massachusetts, Orozco-Muniz, R., Cabrera-Diaz, E. (. (2020). Thermal and chemical treatment of alfalfa seeds (Medicago sativa) and broccoli (Brassica oleracea var. italica) before and during the germination process: An approach to salmonella germination. *Journal of Food Protection*, *83*, 82–88. DOI: 10.4315/0362-028X.JFP-19-272 PMID: 31851547

Hoffman, G. J., Rawlins, S. L., Oster, J. D., Jobes, J. A., & Merrill, S. D. (1979). Leaching requirement for salinity control I. Wheat, sorghum and lettuce. *Agricultural Water Management*, *2*(3), 177–192. DOI: 10.1016/0378-3774(79)90001-5

Inglett, P. W., Reddy, K. R., & Corstanje, R. (2005). *Encyclopedia of soils in the environment.*

IPCC. (2007). *Climate Change 2007: Synthesis Report. Contribution of Working Group I, II and III to the Fourth Assessment Report of the Intergovernmental Panel on Climate Change. Fourth Assessment Report.* Cambridge University Press.

IPCC. (2014). Climate change 2014 synthesis report. *IPCC: Geneva, Szwitzerland*, 1059-1072.

Islam, N., Rashid, M. M., Wibowo, S., Xu, C.-Y., Morshed, A., Wasimi, S. A., Moore, S., & Rahman, S. M. (2021). Early Weed Detection Using Image Processing and Machine Learning Techniques in an Australian Chilli Farm. *Agriculture*, *11*(5), 387. DOI: 10.3390/agriculture11050387

Isleib, J. (2021). Signs and Symptoms of Plant Disease: Is It Fungal, Viral or Bacterial? https://www.canr.msu.edu/news/signs_and_symptoms_of_plant_disease_is_it_fungal_viral_or_bacterial

Jain, N., Bhatia, A., Pathak, H., Gupta, N., Sharma, D. K. and Kaushik, R. (2015). Greenhouse gas emission and global warming. *Introduction to environmental sciences*, 379-411.

Jat, M. L., Sharma, S. K., Cothari, A. K., Jain, L. K., & Balyan, J. K. (2007). Management of drought through crop diversification and intercropping based on rainfall analysis in Western India. *Indian Journal Dryland Agricultural Research Development*, *22*(2), 153–162.

Javed, T., Ali, M.M., Shabbir, R., Gall, S., Ali, A., Khalid, E., Abbas, A.N., Tarik, M. (2020). The emergence of rice seedlings under the influence of various factors and seedling with potassium nitrate. J. *Adj. Res. The science of plants*, 1, 65-75.

Kabato, W., Getnet, G. T., Sinore, T., Nemeth, A., & Molnár, Z. (2025). Towards climate-smart agriculture: Strategies for sustainable agricultural production, food security, and greenhouse gas reduction. *Agronomy (Basel)*, *15*(3), 565. DOI: 10.3390/agronomy15030565

Kadena, M.B., Preston, G.M., Van der Hoorn, R.A., Flanagan, N.C., Townley, O.N., Thompson, I.P. (2018). Increasing the activity of cinnamon essential oil by encapsulating nanoparticles to combat seed pathogens. *Indian crop harvest.* 124, 755-764.

Kaledhonkar, M. J., Meena, B. L., & Sharma, P. C. (2019). Reclamation and nutrient management for salt-affected soils.

Kalendar, R., Orbovic, V., Egea-Cortines, M., & Song, G. (2022). Recent advances in plant genetic engineering and innovative applications. *Frontiers in Plant Science, 13*, 1045417. DOI: 10.3389/fpls.2022.1045417 PMID: 36340337

Kamil, I. A., & Ojoawo, O. J. (2019). Smart irrigation: An ICT application in agriculture. *The International Journal of Engineering and Science, 8*, 53–58.

Kang, M. S. (2024). *Crop improvement: Challenges in the twenty-first century*. CRC Press. DOI: 10.1201/9781003578512

Kanwar, J. S., Bhumbla, D. R., & Sing, N. T. (1965). Studies on the reclamation of saline and sodic soils in the Punjab.

Kefeli, V.I. (1978). Plant growth and natural regulators. *Plant physiology,* 25, 5, Moscow, Science. Al-Mughrabi, K. I. (2007). Effect of treatment of potatoes in storage and pre-planting with hydrogen peroxide (H2O2) on emergence and yield.

Kendouci, M. A., Bendida, A., Khelfaoui, R., & Kharroubi, B. (2013). The impact of traditional irrigation (Foggara) and modern (drip, pivot) on the resource non-renewable groundwater in the Algerian Sahara. *Energy Procedia, 36*, 154–162. DOI: 10.1016/j.egypro.2013.07.018

Khaki, S., & Wang, L. (2019). Crop Yield Prediction Using Deep Neural Networks. *Frontiers in Plant Science, 10*, 621. DOI: 10.3389/fpls.2019.00621 PMID: 31191564

Khan, F., Pandey, P., & Upadhyay, T. K. (2022). Applications of nanotechnology-based agrochemicals in food security and sustainable agriculture: An overview. *Agriculture, 12*(10), 1672. DOI: 10.3390/agriculture12101672

Khan, I., Mesalam, A., Heo, Y. S., Lee, S. H., Nabi, G., & Kong, I. K. (2023). Heat stress as a barrier to successful reproduction and potential alleviation strategies in cattle. *Animals (Basel), 13*(14), 2359. DOI: 10.3390/ani13142359

Khan, S. A., Kumar, S., Hussain, M. Z., & Kalra, N. (2009). Climate change, climate variability and Indian agriculture: impacts vulnerability and adaptation strategies. In *Climate change and crops* (pp. 19–38). Springer Berlin Heidelberg. DOI: 10.1007/978-3-540-88246-6_2

Kirtman, B., Power, S. B., Adedoyin, A. J., Boer, G. J., Bojariu, R., Camilloni, I., Doblas-Reyes, F., Fiore, A. M., Kimoto, M., Meehl, G. and Prather, M. (2013). Near-term climate change: projections and predictability.

Kolhe, P., & Munde, T. N. (2019). Use of drone for efficient water management: A case study. *Proceedings of the 3rd World Irrigation Forum*, 1–13.

Komarov, A.A. (2004). The nature of humic preparations and their effect on cucumber plants, 7, 22 - 23.

Komarov, A.A. (2005). Physiologically active substances reduce the accumulation of nitrates in Chinese cabbage *Potatoes and Vegetables*, 3, 24 - 25.

Kovalenko, L. V., & Folmanis, G. E. (2006). Biologically active iron nanopowders. *Science*, •••, 124.

Kumar, P., Deka, D., Yadav, A., Kumar, M., Das, J. P., Singh, A., & Gurjar, A. (2024). Potential and actual evapotranspiration and Landsat derived indices. *Global Journal of Environmental Science and Management, 10*(3), 1227–1248.

Kumar, V., Kumar, S. S., & Kumar, S. (2016). A study on methane emissions and its mitigation strategies in present scenario. *SAMRIDDHI: A Journal of Physical Sciences. Engineering and Technology, 8*(02), 87–92.

Lallawmkimi, M. C., Veda, D. S., Yadav, A., Kumar, M., & Rout, A. (2024). Innovative approaches in crop genetic engineering for sustainable agriculture: A review. *Journal of Advances in Biology & Biotechnology, 27*(8), 615–631. DOI: 10.9734/jabb/2024/v27i81177

Lal, M. (2003). Global climate change: India's monsoon and its variability. *Journal of Environmental Studies and Policy, 6*(1), 1–34.

Lal, R. (2004). Soil carbon sequestration in India. *Climatic Change, 65*(3), 277–296. DOI: 10.1023/B:CLIM.0000038202.46720.37

Lal, R. (2012). Climate change and soil degradation mitigation by sustainable management of soils and other natural resources. *Agricultural Research, 1*(3), 199–212. DOI: 10.1007/s40003-012-0031-9

Lal, R. (2013). Intensive agriculture and the soil carbon pool. *Journal of Crop Improvement, 27*(6), 735–751. DOI: 10.1080/15427528.2013.845053

Lampridi, M., Sørensen, C., & Bochtis, D. (2019). Agricultural Sustainability: A Review of Concepts and Methods. *Sustainability (Basel), 11*(18), 5120. DOI: 10.3390/su11185120

LeCun, Y., Bengio, Y., & Hinton, G. (2015). Deep learning. *Nature, 521*(7553), 436–444. DOI: 10.1038/nature14539 PMID: 26017442

Liakos, K., Busato, P., Moshou, D., Pearson, S., & Bochtis, D. (2018). Machine Learning in Agriculture: A Review. *Sensors (Basel), 18*(8), 2674. DOI: 10.3390/s18082674 PMID: 30110960

Li, N., Ren, Z., Li, D., & Zeng, L. (2020). Review: Automated techniques for monitoring the behavior and welfare of broilers and laying hens: Towards the goal of precision livestock farming. *Animal, 14*(3), 617–625. DOI: 10.1017/S1751731119002155 PMID: 31566170

Liu, Z., & Xu, Q. (2018). An automatic irrigation control system for soilless culture of lettuce. *Water (Basel), 10*(11), 1692. DOI: 10.3390/w10111692

Lugovskoy potatoes: subtleties, yield, cultivation and care URL: https://dzen.ru/a/YlVl2BQqfmgppBSK. Date of acessed 27/08/2024

Lu, Y., Xu, W., Leng, J., Liu, X., Xu, H., Ding, H., Zhou, J., & Cui, L. (2024). Review and Research Prospects on Additive Manufacturing Technology for Agricultural Manufacturing. *Agriculture, 14*(8), 1207. DOI: 10.3390/agriculture14081207

Lybbert, T. and Sumner, D. (2010). Agricultural technologies for climate change mitigation and adaptation in developing countries: policy options for innovation and technology diffusion.

Maas, E. V., & Hoffman, G. J. (1977). Crop salt tolerance—Current assessment. *Journal of the Irrigation and Drainage Division, 103*(2), 115–134. DOI: 10.1061/JRCEA4.0001137

Malhi, G. S., Kaur, M., & Kaushik, P. (2021). Impact of climate change on agriculture and its mitigation strategies: A review. *Sustainability (Basel), 13*(3), 1318. DOI: 10.3390/su13031318

Malik, L., Sanaullah, M., Mahmood, F., Hussain, S., Siddique, M. H., Anwar, F., & Shahzad, T. (2022). Unlocking the potential of co-applied biochar and plant growth-promoting rhizobacteria (PGPR) for sustainable agriculture under stress conditions. *Chemical and Biological Technologies in Agriculture, 9*(1), 58. DOI: 10.1186/s40538-022-00327-x PMID: 37520585

Malviya, A., & Jaspal, D. (2021). Artificial intelligence as an upcoming technology in wastewater treatment: A comprehensive review. *Environmental Technology Reviews, 10*(1), 177–187. DOI: 10.1080/21622515.2021.1913242

Mamata, K., & Lamichhane, A. (2021). Advances in agricultural biotechnology. *Nepal Journal of Biotechnology, 9*(1), 85–92. DOI: 10.3126/njb.v9i1.38643

Mannion, A., & Morse, S. (2013). GM crops 1996–2012: A review of agronomic, environmental and socio-economic impacts. *University of Surrey, Centre for Environmental Strategy Working Paper, 4*(13), 1–40.

Martínez, R. T., Cruz-Martinez, Y. R., Villagómez-Guzmán, A. K., Pérez-Valera, O., Arreaga-González, H. M., & Hernández-Delgado, T. (2025). Biofertilizers and Biopesticides: Sustainable Alternatives for Agriculture. In *Soil Improvement and Water Conservation Biotechnology* (pp. 113–136). Bentham Science Publishers.

Matis, J. H., Birket, T., & Boudreaux, D. (1989). An Application of the Markov Chain Approach to Forecasting Cotton Yields from Surveys. *Agricultural Systems, 29*(4), 357–370. DOI: 10.1016/0308-521X(89)90097-8

Mehraj, S., Manzoor, M. M. M., Sharma, R. K., Mir, A. H., Khan, I. L., & Maqbool, S. (2022). Climate change and agriculture. In *Environmental Studies and Climate Change* (pp. 231–239). CRC Press. DOI: 10.1201/9781003220824-17

Melnyk, L., Kalinichenko, L., Kubatko, O., Dobrowolski, Z., & Babczuk, A. (2023). *Restructuring the economic systems on the way to an additive economy.*

Miller, L., Vellidis, G., Mohawesh, O., & Coolong, T. (2018). Comparing a smartphone irrigation scheduling application with water balance and soil moisture-based irrigation methods: Part I—Plasticulture-grown tomato. *HortTechnology, 28*(3), 354–361. DOI: 10.21273/HORTTECH04010-18

Minhas, P. S., & Gupta, R. K. (1992). *Quality of irrigation water: assessment and management.* Indian Council of Agricultural Research.

Ministry of Agriculture and Farmers Welfare. (2014). *National Mission on Sustainable Agriculture.* NMSA.

Ministry of Agriculture and Farmers Welfare. (2015). Soil Health Card Scheme. https://soilhealth.dac.gov.in/home

Ministry of Agriculture and Farmers Welfare. (2016). Paramparagat Krishi Vikas Yojana (PKVY). https://nmsa.dac.gov.in/ pdfDoc/SHM_Guidelines472016.pdf

Ministry of Agriculture and Farmers Welfare. (2016). Pradhan Mantri Fasal Bima Yojana (PMFBY). https://www.pmfby.gov.in

Ministry of Agriculture and Farmers Welfare. (2016). *Weather based Crop Insurance Scheme*. WBCIS.

Ministry of Jal Shakti. (2015). Pradhan Mantri Krishi Sinchayee Yojana (PMKSY). https://pmksy-mowr.nic.in

Mochida, O. M., Joshi, R. C. and Litsinger, J. A. (1987). Climatic factors affecting the occurrence of insect pests. *Weather and rice*, 149-164.

Mohammad, F. S., Al-Ghobari, H. M., & El Marazky, M. S. A. (2013). Adoption of an intelligent irrigation scheduling technique and its effect on water use efficiency for tomato crops in arid regions. *Australian Journal of Crop Science*, *7*(3), 305–313.

Monisha Kaul, M., Robert, L., Hill, H., & Walthall, C. (2005). Artificial neural networks for corn and Soybean yield prediction, *Elsevier. Agricultural Systems*, *85*(1), 1–18. DOI: 10.1016/j.agsy.2004.07.009

Morales, L., Domínguez, M. T., & Fernández-Boy, E. (2022, May). Effect of the addition of organic amendments to C-poor agricultural soils on soil resistance against drought. In EGU General Assembly Conference Abstracts (pp. EGU22-5532).

Munasinghe, M., Canziani, O., Davidson, O., Metz, B., Parry, M., & Harisson, M. (2003). *Integrating sustainable development and climate change in the IPCC Fourth Assessment Report*. Munasinghe Institute for Development.

Mustafa, , GShehzad, , M. ATahir, , M. H. NNawaz, , FAkhtar, , GBashir, , M. AGhaffar, , A. (2022). Pretreatment with chitosan arbitrates physiological processes and antioxidant defense system to increase drought tolerance in alfalfa (Medicago sativa L.). *Journal of Soil Science and Plant Nutrition, 22*(2), 2169-2186.

Nassani, A. A., Awan, U., Zaman, K., Hyder, S., Aldakhil, A. M., & Abro, M. M. Q. (2019). Management of natural resources and material pricing: Global evidence. *Resources Policy*, *64*, 101500. DOI: 10.1016/j.resourpol.2019.101500

Navarro-Hellin, H., Rincon, M. D., Martinez-del-Rincon, J., Sotos, V. F., & Torres, S. R. (2016). A decision support system for managing irrigation in agriculture. *Computers and Electronics in Agriculture*, *124*, 121–131. DOI: 10.1016/j.compag.2016.04.003

Nema, P., Nema, S., & Roy, P. (2012). An overview of global climate changing in current scenario and mitigation action. *Renewable & Sustainable Energy Reviews*, *16*(4), 2329–2336. DOI: 10.1016/j.rser.2012.01.044

Neupane, J., & Guo, W. (2019). Agronomic Basis and Strategies for Precision Water Management: A Review. *Agronomy (Basel)*, *9*(2), 87. DOI: 10.3390/agronomy9020087

Neven, L. G. (2000). Physiological responses of insects to heat. *Postharvest Biology and Technology*, *21*(1), 103–111. DOI: 10.1016/S0925-5214(00)00169-1

Nielsen, D. R., Biggar, J. W., & Luthin, J. N. (1966). Desalinization of soils under controlled unsaturated flow conditions. In *Proceedings of the 6th International Congress on Irrigation and Drainage* (pp. 15-19).

Norikane, J. H., Brook, R. C., & Kirk, W. W. (2001). Efficacy of purogene and oxidate disinfectants added to potato storage humidity water for pathogen control. *American Journal of Potato Research*, *78*, 473–474.

Novak, S. M., & Fiorelli, J. L. (2010). Greenhouse gases and ammonia emissions from organic mixed crop-dairy systems: A critical review of mitigation options. *Agronomy for Sustainable Development*, *30*(2), 215–236. DOI: 10.1051/agro/2009031

Nwanze, K. F., & Fan, S. (2016). Climate change and agriculture: strengthening the role of smallholders. In *Global food policy report* (pp. 13–21). International Food Policy Research Institute.

O'Mara, F. P. (2012). The role of grasslands in food security and climate change. *Annals of Botany*, *110*(6), 1263–1270. DOI: 10.1093/aob/mcs209

Oerke, E. C. (2006). Crop losses to pests. *Journal of Agricultural Science*, *144*(1), 31–43. DOI: 10.1017/S0021859605005708

Olayide, O. E., Tetteh, I. K., & Popoola, L. (2016). Differential impacts of rainfall and irrigation on agricultural production in Nigeria: Any lessons for climate-smart agriculture. *Agricultural Water Management, 178*, 30–36. DOI: 10.1016/j.agwat.2016.08.034

Olesen, J. E. (2016). Socio-economic impacts—Agricultural systems. *North Sea Region Climate Change Assessment*, 397–407.

Olhoff, A., & Schaer, C. (2010). *Screening Tools and Guidelines to Support the Mainstreaming of Climate Change Adaptation into Development Assistance – A Stocktaking Report*. UNDP.

Olsen, N., Kleinkopf, G., Secor, G., Woodell, L., & Nolte, P. (2000). The use of chlorine dioxide in potato storage. *Bulletin 825,* University of Idaho, 4.

Osman, K. T. (2018). *Management of Soil Problems*. Springer. DOI: 10.1007/978-3-319-75527-4

Pal, L., Ojha, C. S. P., & Dimri, A. P. (2021). Characterizing rainfall occurrence in India: Natural variability and recent trends. *Journal of Hydrology (Amsterdam), 603*, 126979. DOI: 10.1016/j.jhydrol.2021.126979

Panda, S. K., Baluška, F., & Matsumoto, H. (2009). Aluminum stress signaling in plants. *Plant Signaling & Behavior, 4*(7), 592–597. DOI: 10.4161/psb.4.7.8903

Pandey, D. N., Gupta, A. K., & Anderson, D. M. (2003). Rainwater harvesting as an adaptation to climate change. *Current Science*, •••, 46–59.

Pandiselvam, R., Mayookha, V. P., Kothakota, A., Sharmila, L., Ramesh, S. V., Bharathi, C. P., Gomathy, K., & Srikanth, V. (2020). Ozone treatment method for seed germination - a systematic review. *Ozone Science and Engineering, 42*(4), 331–346. DOI: 10.1080/01919512.2019.1673697

Pandiyan, A., Sarsan, S., Durga, G. G. S., & Ravikumar, H. (2024). Biofertilizers and biopesticides as microbial inoculants in integrated pest management for sustainable agriculture. In *Microbial essentialism* (pp. 485–518). Elsevier. DOI: 10.1016/B978-0-443-13932-1.00010-6

Pangga, I. B., Hanan, J., & Chakraborty, S. (2013). Climate change impacts on plant canopy architecture: Implications for pest and pathogen management. *European Journal of Plant Pathology, 135*(3), 595–610. DOI: 10.1007/s10658-012-0118-y

Panhwar, Q. A., Radziah, O., Zaharah, A. R., Sariah, M., & Mohd Razi, I. (2012). Isolation and characterization of phosphorus solubilizing bacteria. *African Journal of Biotechnology, 11*, 2711–2719.

Pan, S. J., & Yang, Q. (2010). A survey on transfer learning. *IEEE Transactions on Knowledge and Data Engineering, 22*(10), 1345–1359. DOI: 10.1109/TKDE.2009.191

Papageorgiou, E. I., Aggelopoulou, K., Gemtos, T. A., & Nanos, G. D. (2018). Development and Evaluation of a Fuzzy Inference System and a Neuro-Fuzzy Inference System for Grading Apple Quality. *Applied Artificial Intelligence, 32*(3), 253–280. DOI: 10.1080/08839514.2018.1448072

Parmar, P., Kumar, R., Neha, Y., & Srivatsan, V. (2023). Microalgae as next generation plant growth additives: Functions, applications, challenges and circular bioeconomy based solutions. *Frontiers in Plant Science, 14*, 1073546. DOI: 10.3389/fpls.2023.1073546 PMID: 37063190

Pasko O.A. (1991). Acceleration of seed germination when treated with electrochemically activated water. *Physiology and biochemistry of cultivated plants*, 23 (6), 552 – 556.

Pasko, O. A. (2000). *Activated water and its use in agriculture: monograph*. Tomsk: TPU, 180.

Pasko, O. A. (2000a). *Ecological aspects of increasing the productivity of flower, vegetable crops and potatoes in the taiga zone of Western Siberia*: abstract of the dissertation... Doctor of Agricultural Sciences: specialty 06.01.09, 03.00.16. Novosibirsk State Agrarian University. Univ.: Novosibirsk, 32.

Pasko, O. A. (2009). Pat. 2375313 RU, IPC C 02 F 1/461. *Flow-through diaphragm electrolyzer* No. 2007118618/15. Published on 10.12.2009.

Pasko, O. A. (2010b). Stimulation of vegetative reproduction of ornamental plants using electrochemically activated water. *Phytodesign in modern conditions materials of the International Scientific and Practical Conference*, Belgorod: BelSU Publishing House, 146-147.

Pasko, O. A. (2024). *Method of obtaining stimulator of growth.* Retrieved from URL https://patents.google.com/patent/SU1574196A1/en Date of acessed 27/08/2024

Pasko, O.A. & Leonov V. P. (2011) The mechanism of action of activated water on rhizogenesis Cherenkov. The mechanism of activated water on rhizogenesis cuts. *Modern problems of science and education*, 5, 10.

Pasko, O.A. & Semenov, A.V. (1991). *Method of pre-sowing seed treatment.* Copyright certificate No. 1667674 of the USSR, class A 01 C 7 /00. Publ. 07.06.1991. Byul. No. 29.

Pasko, O.A. (1991). Acceleration of seed germination when treated with electrochemically activated water. *Physiology and biochemistry of cultivated plants*, 23 (6), 552 – 556.

Pasko, O.A. (2010a). The growth and development of plants stimulated by electrochemically activated water. *Bulletin of the B.S.S.A. named after V.R. Filippov scientific and theoretical journal*, 3, 54-59.

Pasko, O.A. (2010a). The growth and development of plants stimulated by electrochemically activated water. *Bulletin of the B.S.S.A. named after V.R. Filippov scientific and theoretical journal*, 3,54-59.

Pasko, O.A. (2011a). Stimulation of tomato seeds with activated water increases the yield. *Potatoes and vegetables*, 6, 21.

Pasko, O.A., Semenov, A.V. (1993a), *A method for obtaining a plant growth and development stimulant.* Copyright certificate No. 161 19557 USSR, class A 01 N 59/00. Publ. 06/07/1993. Bul. No. 21.

Pasko, O.A., Semenov, A.V. (1993b) *A method for obtaining a plant growth and development stimulant.* Copyright certificate No. 16119557 of the USSR, class A 01 N 59/00.

Pasko, O. A. (2010c). Physico-chemical changes in tap water during its treatment in various ways. Property changes of piped water treated with different methods. *Water: Chemistry and Ecology*, *7*, 40–45.

Pasko, O. A. (2011). The influence of pre-sowing stimulation of cucumber seeds on yield. *Agricultural Sciences*, *8*, 20–22.

Pasko, O. A. (2011b). Stimulation of vegetative reproduction of plants with sodium humate Innovative aspects of peat extraction, processing and application. *Proceedings of the International Conference dedicated to the 115th anniversary of the National Research* Tomsk Polytechnic University, Tomsk: TPU Publishing House, 122-125.

Pasko, O. A. (2011b). The effect of presowing stimulation of cucumber seeds on yield. Influence of presowing stimulation of cucumber seeds on productivity. Agricultural science. Scientific. *Theoretical and Industrial Journal*, *8*, 20–22.

Pasko, O. A., & Gomboev, D. D. (2011). *Activated water and the possibilities of its application in crop production and animal husbandry*. Tomsk Polytechnic University Publishing House.

Pasko, O. A., & Ketskalo, V. M. (1991). The influence of electrochemically activated water on seed germination. *Plant Physiology*, *38*(2), 381–385.

Pasko, O. A., Semenov, A. V., Smirnov, G. V., & Smirnov, D. G. (2007). *Activated liquids, electromagnetic fields and flicker noise. Their application in medicine and agriculture*. TUSUR Publishing House.

Pathak, H., Aggarwal, P. K., & Singh, S. D. (2012). Climate change impact, adaptation and mitigation in agriculture: methodology for assessment and applications. *Indian Agricultural Research Institute, New Delhi*, 302, 19.

Pathak, H. (2023). Impact, adaptation, and mitigation of climate change in Indian agriculture. *Environmental Monitoring and Assessment*, *195*(1), 52. DOI: 10.1007/s10661-022-10537-3

Pathak, H., Byjesh, K., Chakrabarti, B., & Aggarwal, P. K. (2011). Potential and cost of carbon sequestration in Indian agriculture: Estimates from long-term field experiments. *Field Crops Research, 120*(1), 102–111. DOI: 10.1016/j.fcr.2010.09.006

Pathak, H., Ladha, J. K., Aggarwal, P. K., Peng, S., Das, S., Singh, Y., Singh, B., Kamra, S. K., Mishra, B., Sastri, A. S. R. A. S., Aggarwal, H. P., Das, D. K., & Gupta, R. K. (2003). Trends of climatic potential and on-farm yields of rice and wheat in the Indo-Gangetic Plains. *Field Crops Research, 80*(3), 223–234. DOI: 10.1016/S0378-4290(02)00194-6

Pathak, R., Singh, S. K., Tak, A., & Gehlot, P. (2018). Impact of climate change on host, pathogen and plant disease adaptation regime: A review. *Biosciences Biotechnology Research Asia, 15*(3), 529–540. DOI: 10.13005/bbra/2658

Peters, J. B., Schulte, E. E., & Kelling, K. A. (1996). Choosing between liming materials. University of Wisconsin--Extension.

Pingali, P., Aiyar, A., Abraham, M., Rahman, A., Pingali, P., Aiyar, A., Abraham, M., & Rahman, A. (2019). Diet diversity and the declining importance of staple grains. *Transforming food systems for a rising India*, pp.73-91.

Potato variety Lugovskoy: characteristics and description, reviews, photos. URL: https://fermilon.ru/sad-i-ogorod/ovoshhi/sort-kartofelya-lugovskoy-harakteristika-i-opisanie-otzyvy-foto.html. Date of access 08/27/24.

Potato variety Nevsky. State variety testing. URL: https://gossortrf.ru/registry/gosudarstvennyy-reestr-selektsionnykh-dostizheniy-dopushchennykh-k-ispolzovaniyu-tom-1-sorta-rasteni/nevskiy-kartofel. Date of access 08/27/24.

Potato variety Nevsky: features, planting and care. URL: https://ferma.expert/rasteniya/ovoshchi/kartofel/kartofel-nevskij. Date of access 08/27/24.

Powell, B., & Martens, M. (2005). A review of acid sulfate soil impacts, actions and policies that impact on water quality in Great Barrier Reef catchments, including a case study on remediation at East Trinity. *Marine Pollution Bulletin*, *51*(1-4), 149–164. DOI: 10.1016/j.marpolbul.2004.10.047

Prasad, R. S., Mohapatra, T., & Rao, S. (2019). *Climate change and Indian agriculture: impacts, coping strategies, programmes and policy*. Indian Council of Agricultural Research.

Puttaswamygowda, B. S., & Pratt, P. F. (1973). Effects of straw, calcium chloride and submergence on a sodic soil. *Soil Science Society of America Journal*, *37*(2), 208–212. DOI: 10.2136/sssaj1973.03615995003700020016x

Qadir, M., Oster, J. D., Schubert, S., Noble, A. D., & Sahrawat, K. L. (2007). Phytoremediation of sodic and saline-sodic soils. *Advances in Agronomy*, *96*, 197–247. DOI: 10.1016/S0065-2113(07)96006-X

Quaggio, J. A., Van Raij, B., & Malavolta, E. (1985). Alternative use of the SMP-buffer solution to determine lime requirement of soils. *Communications in Soil Science and Plant Analysis*, *16*(3), 245–260. DOI: 10.1080/00103628509367600

Rahman, M. A., Lee, S. H., Ji, H. C., Kabir, A. H., Jones, C. S., & Lee, K. W. (2018). Importance of mineral nutrition for mitigating aluminum toxicity in plants on acidic soils: Current status and opportunities. *International Journal of Molecular Sciences*, *19*(10), 3073. DOI: 10.3390/ijms19103073

Rahman, M. M., Uddin, S., Jahangir, M. M. R., Solaiman, Z. M., Alamri, S., Siddiqui, M. H., & Islam, M. R. (2021). Integrated nutrient management enhances productivity and nitrogen use efficiency of crops in acidic and charland soils. *Plants*, *10*(11), 2547. DOI: 10.3390/plants10112547

Raihan, A. (2024). A review of recent advances, challenges and potential future directions of climate-smart agriculture. In *Proceedings of The International Conference on Climate-Smart Agriculture*.

Ramanathan, V., & Feng, Y. (2009). Air pollution, greenhouse gases and climate change: Global and regional perspectives. *Atmospheric Environment, 43*(1), 37–50. DOI: 10.1016/j.atmosenv.2008.09.063

Ramesh, R., Negi, S. C., & Rana, S. S. (2016). Resource conservation technologies (RCTs)-needs and future prospects. *RE:view*.

Rao, B. B., Chowdary, P. S., Sandeep, V. M., Rao, V. U. M., & Venkateswarlu, B. (2014). Rising minimum temperature trends over India in recent decades: Implications for agricultural production. *Global and Planetary Change, 117*, 1–8.

Rao, C. S., Baral, K., Chanadana, V. M., Jagadesh, M., & Karthik, R.V. MANI CHANADANAM. JAGADESHR. KARTHIK. (2024). Climate change adaptation and mitigation in Indian agriculture. *Journal of Agrometeorology, 26*(2), 137–148. DOI: 10.54386/jam.v26i2.2582

Rashad, S. (2021). An overview on the aquatic fern Azolla spp. as a sustainable source of nutrients and bioactive compounds with resourceful applications. *Egyptian Journal of Aquatic Biology and Fisheries, 25*(1), 775–782. DOI: 10.21608/ejabf.2021.150205

Ray, S. K., Chatterjee, D., Saha, S., Kamble, K. H., Deka, B. C., Kendra, V., & Jharnapani, M. (2015). *A guide for soil nutrient management with special reference to Wokha. ICAR Research Complex for NEH Region.* Nagaland Centre.

Reddy, P. P. and Reddy, P. P. (2015). Climate change adaptation. *Climate resilient agriculture for ensuring food security*, 223-272.

Reeve, R. C., Pillsbury, A. F., & Wilcox, L. V. (1955). Reclamation of a saline and high boron soil in the Coachella Valley of California. Hilgardia;(United States), 24(4). DOI: 10.3733/hilg.v24n04p069

Rhoades, J. D., & Loveday, J. (1990). Salinity in irrigated agriculture. *Agronomy (Basel)*, (30), 1089–1142.

Risser, M. D., Collins, W. D., Wehner, M. F., O'Brien, T. A., Huang, H., & Ullrich, P. A. (2024). Anthropogenic aerosols mask increases in US rainfall by greenhouse gases. *Nature Communications*, *15*(1), 1318. DOI: 10.1038/s41467-024-45504-8

Robbins, C. W., & Gavlak, R. G. (1989). Salt-and sodium-affected soils.

Robinet, C., & Roques, A. (2010). Direct impacts of recent climate warming on insect populations. *Integrative Zoology*, *5*(2), 132–142. DOI: 10.1111/j.1749-4877.2010.00196.x

Rockström, J., Karlberg, L., Wani, S. P., Barron, J., Hatibu, N., Oweis, T., Bruggeman, A., Farahani, J., & Qiang, Z. (2010). Managing water in rainfed agriculture—The need for a paradigm shift. *Agricultural Water Management*, *97*(4), 543–550. DOI: 10.1016/j.agwat.2009.09.009

Rodríguez, J. Á. C., Hanafi, M. M., Omar, S. R. S., & Rafii, Y. M. (2009). Chemical characteristics of representative high aluminium saturation soil as affected by addition of soil amendments in a closed incubation system. *The Malaysian Journal of Soil Science*, *13*, 13–28.

Rupa Kumar, K., Ashrit, R. G., & Pant, G. B. (2002). The Indian summer monsoon: Past, present and future. *Science and Culture*, *68*(9-12), 217–224.

Ruzinazarov, S., Achilova, L., Ismailova, R., & Men, V. (2024). Digitalization and legislative support in intellectual property protection for fodder additive production technologies. *BIO Web of Conferences, 149*, 01010.

Rybinsky, V. (2001). The effect of laser beams in combination with chemomutagen (MNU) on the variability of traits and mutation frequency of spring barley. *International Agrophysics*, *15*, 115–119.

Saab, M. T., Jomma, I., Skaf, S., & Fahed, S. (2019). Assessment of a smartphone application for real-time irrigation scheduling in Mediterranean environments. *Water (Basel)*, *11*(2), 2–18.

Sajjad, S. H. 2008. Impacts of climate change on south asia.

Sakala, E., & Demchuk, A., Grzhi, Oe E., Proeba-Bialchik, U., Scheisner, H. (2012). The effect of pre-sowing laser irradiation of seeds on the properties of sugar beet. *International Agrophysics*, *26*, 295–300. DOI: 10.2478/v10247-012-0042-6

Samarah, , N. HAl-Quraan, , N. AMassad, , R. SWelbaum, , G. E. (2020). Treatment of bell pepper (Capsicum annuum L.) seeds with chitosan increases chitinase and glucanase activities and enhances emergence in a standard cold test. *Scientia Horticulturae, 269*, 109393.

Samuel, A. L. (1959). Some Studies in Machine Learning Using the Game of Checkers. *IBM Journal of Research and Development*, *44*(1.2), 206–226. DOI: 10.1147/rd.441.0206

Sandoval, F. M., & Benz, L. C. (1966). Effect of bare fallow, barley and grass on salinity of a soil over a saline water table. *Soil Science Society of America Journal*, *30*(3), 392–396. DOI: 10.2136/sssaj1966.03615995003000030026x

Sapkota, T. B., Jat, M. L., Aryal, J. P., Jat, R. K., & Khatri-Chhetri, A. (2015). Climate change adaptation, greenhouse gas mitigation and economic profitability of conservation agriculture: Some examples from cereal systems of Indo-Gangetic Plains. *Journal of Integrative Agriculture*, *14*(8), 1524–1533. DOI: 10.1016/S2095-3119(15)61093-0

Sarangi, S. K., Maji, B., Digar, S., Burman, D., Mandal, U. K., Mahanta, K. K., & Mandal, S. (2019). Technologies for Cropping System Intensification in the Salt-Affected Coastal Zone of West Bengal, India Technical Bulletin ICAR-CSSRI, p. 28

Sarangi, S. K., & Maji, B. (2017). Sustainable rice cultivation in coastal saline soils: a case study Institute, India. In *Achieving sustainable cultivation of rice* (Vol. 2, pp. 101–136). Burleigh Dodds Science Publishing. DOI: 10.19103/AS.2016.0003.07

Sarkar, A. K., Pattanayak, S. K., Singh, S., Mahapatra, P., Kumar, A., & Ghosh, G. K. (2020). Integrated nutrient management strategies for acidic soils. *Indian Journal of Fertilisers*, *16*(5), 476–491.

Secor, G. A., & Gudmestad, N. C. (1999). Managing fungal diseases of potato. *Canadian Journal of Plant Pathology*, *21*(3), 213–221. DOI: 10.1080/07060669909501184

Seeland, M., Rzanny, M., Alaqraa, N., Wäldchen, J., & Mäder, P. (2017). Plant species classification using flower images—A comparative study of local feature representations. *PLoS One*, *12*(2), e0170629. DOI: 10.1371/journal.pone.0170629 PMID: 28234999

Sehgal, G., Gupta, B., Paneri, K., Singh, K., Sharma, G., & Shroff, G. (2017). Crop Planning using Stochastic Visual Optimization. *arXiv preprint arXiv:1710.09077*. DOI: 10.1109/VDS.2017.8573443

Semenenko, S. Ya., Belitskaya, M. N., & Likholetov, S. M. (2013). Phytosanitary improvement of grain and vegetable crops using electrochemically activated water // Advances in modern natural science. 1, 78-82. URL: https://natural-sciences.ru/ru/article/view?id=31286 (date of access: 08/27/2024).

Senthil, K. S. D., & Shamili, M. D. (2022).Smart farming using Machine Learning and Deep Learning techniques, *Decision Analytics Journal,* Vol. 3,ISSN 2772-6622,DOI: 10.1016/j.dajour.2022.100041

Shaikh, T. N., & Mukane, S. M. (2017). Segmentation of Pomegranate Leaf for Detection of Disease Using Image Processing. *International Journal of Innovative Research in Computer and Communication Engineering*, *5*(4), 378–384.

Shainberg, I., & Shalhevet, J. (Eds.). (2012). *Soil salinity under irrigation: Processes and management* (Vol. 51). Springer Science & Business Media.

Shamshuddin, J., Panhwar, Q. A., Shazana, M. A. R. S., Elisa, A. A., Fauziah, C. I., & Naher, U. A. (2016). Improving the productivity of acid sulfate soils for rice cultivation using limestone, basalt, organic fertilizer and/or their combinations. *Sains Malaysiana*, *45*(3), 383–392.

Shanker, A. K., Gunnapaneni, D., Bhanu, D., Vanaja, M., Lakshmi, N. J., Yadav, S. K., Prabhakar, M., & Singh, V. K. (2022). Elevated CO_2 and water stress in combination in plants: Brothers in arms or partners in crime. *Biology (Basel)*, *11*(9), 1330. DOI: 10.3390/biology11091330

Sharma, D. R. (2008). *Characterization of Sodic Water and Technologies for its Efficient Use*. Conjunctive Use of Canal and Groundwater.

Sharma, K. D., & Singh, P. (2007). Impact of climate change on hydrological extremes: Floods and droughts. *Hydrology Journal*, *30*(34), 129–145.

Shera, B., Scholz, V., Ireshova, Yu., Hong, J., Yulak, J., & Sheri, M. (2021). Investigation of non—Thermal plasma treatment on seed germination and early growth of legumes - review. *Plants*, *10*, 1616. DOI: 10.3390/plants10081616 PMID: 34451662

Siebert, S., Burke, J., Faures, J. M., Frenken, K., Hoogeveen, J., Döll, P., & Portmann, F. T. (2010). Groundwater use for irrigation–a global inventory. *Hydrology and Earth System Sciences*, *14*(10), 1863–1880. DOI: 10.5194/hess-14-1863-2010

Singh, J., Sahany, S., Singh, K. K., Robock, A. and Xia, L. (2024). Future climate change impacts on rice in Uttar Pradesh, India's most populous agrarian state. *Earth's Future*, 12(5), e2023EF004009.

Singh, K. M., Kumar, A., & Singh, R. K. P. (2017). Role of information and communication technologies in Indian agriculture: An overview. SSRN *Electronic Journal*, 1–12.

Singh, R. P., Kumar, S., Sainger, M., Sainger, P. A. and Barnawal, D. (2017). Eco-friendly nitrogen fertilizers for sustainable agriculture. *Adaptive soil management: From theory to practices*, 227-246.

Singh, Y. P., Sharma, D. K., Singh, G., Nayak, A. K., & Mishra, V. K. (2008). Alternate Land Use Management for Sodic Soils. Technical bulletin no. 2/2008.

Singh, B. K., Delgado-Baquerizo, M., Egidi, E., Guirado, E., Leach, J. E., Liu, H., & Trivedi, P. (2023). Climate change impacts on plant pathogens, food security and paths forward. *Nature Reviews. Microbiology*, *21*(10), 640–656. DOI: 10.1038/s41579-023-00900-7

Singh, B., Singh, V., Purba, J., Sharma, R. K., Jat, M. L., Singh, Y., Thind, H. S., Gupta, R. K., Chaudhary, O. P., Chandna, P., Khurana, H. S., Kumar, A., Singh, J., Uppal, H. S., Vashistha, M., & Gupta, R. (2015). Site-specific fertilizer nitrogen management in irrigated transplanted rice (*Oryza sativa*) using an optical sensor. *Precision Agriculture*, *16*(4), 455–475. DOI: 10.1007/s11119-015-9389-6

Singh, J., Heeren, D. M., Ge, Y., Geng, B., Neale, C. M. U., Maguire, M. S., & Bhatti, S. (2021). Sensor-based irrigation of maize and soybean in east-central Nebraska under a sub-humid climate. *Biological Systems Engineering: Papers and Publications*, *763*, 1–12. DOI: 10.13031/aim.202101044

Singh, S., Ashoka, P., Ahlawat, U., Changdeo, W. B., Rehsawla, R., Naruka, A., & Sharma, D. (2024). Mechanisms and Applications of Microbial Biotechnology in Soil Health and Agricultural Productivity: A Review. *Journal of Advances in Biology & Biotechnology*, *27*(7), 1420–1438. DOI: 10.9734/jabb/2024/v27i71104

Sinwar, D., Dhaka, V. S., Sharma, M. K., & Rani, G. (2019). AI-based yield prediction and smart irrigation. *Springer Nature*, *2*, 155–179.

Sivakumar, M. V. (2007). Interactions between climate and desertification. *Agricultural and Forest Meteorology*, *142*(2-4), 143–155. DOI: 10.1016/j.agrformet.2006.03.025

Sivasubramaniam, K., Gita, R., Sujata, K., Raja, K., Sripunitha, A., & Selvarani, R. (2011). Seed sowing: Triumphs and aDWersities. *Madras is. Agricultural Journal*, *98*, 197–209.

Skendžić, S., Zovko, M., Živković, I. P., Lešić, V., & Lemić, D. (2021). The impact of climate change on agricultural insect pests. *Insects*, *12*(5), 440. DOI: 10.3390/insects12050440

Slafer, G. A., Molina-Cano, J. L., Savin, R., Araus, J. L., & Romagosa, I. (2024). *Barley science: Recent advances from molecular biology to agronomy of yield and quality.*

Slaughter, D. C., Giles, D. K., & Downey, D. (2008). Autonomous robotic weed control systems: A review. *Computers and Electronics in Agriculture*, *61*(1), 63–78. DOI: 10.1016/j.compag.2007.05.008

Smilovic, M., Gleeson, T., & Siebert, S. (2015). The limits of increasing food production with irrigation in India. *Food Security*, *7*(4), 835–856. DOI: 10.1007/s12571-015-0477-2

Smith, A. (2025). Systemic Analysis of Circular Economy Principles in Agricultural Waste Management and Their Synergy with National Sustainable Development Objectives. *Transdisciplinary Advances in Social Computing, Complex Dynamics, and Computational Creativity*, *15*(2), 10–21.

Smola, A. (1999). Regression Estimation with Support Vector Learning Machines. *Master's Thesis, The Technical University of Munich, Munich, Germany*, pp. 1–78.

Sohail, M., Pirzada, T., Opperman, C. H., & Khan, S. A. (2022). Recent advances in seed coating technologies: Transitioning toward sustainable agriculture. *Green Chemistry*, *24*(16), 6052–6085. DOI: 10.1039/D2GC02389J

Somasundaram, J., Sinha, N. K., Dalal, R. C., Lal, R., Mohanty, M., Naorem, A. K., Hati, K. M., Chaudhary, R. S., Biswas, A. K., Patra, A. K., & Chaudhari, S. K. (2020). No-till farming and conservation agriculture in South Asia–issues, challenges, prospects and benefits. *Critical Reviews in Plant Sciences*, *39*(3), 236–279. DOI: 10.1080/07352689.2020.1782069

Soni, R., Gupta, R., Agarwal, P., & Mishra, R. (2022). Organic farming: A sustainable agricultural practice. *Vantage: Journal of Thematic Analysis*, *3*(1), 21–44. DOI: 10.52253/vjta.2022.v03i01.03

Sonka, S. (2016). Big Data: Fueling the Next Evolution of Agricultural Innovation. *J. Innov. Manag.*, *4*(1), 114–136. DOI: 10.24840/2183-0606_004.001_0008

Sørensen, C. A. G., Kateris, D., & Bochtis, D. (2019). ICT Innovations and Smart Farming. In *Communications in Computer and information science* (Vol. 953, pp. 1–19). Springer.

Squire, H. J., Tomatz, S., Voke, E., González-Grandío, E., & Landry, M. (2023a). The emerging role of nanotechnology in plant genetic engineering. *Nature Reviews Bioengineering*, *1*(5), 314–328. DOI: 10.1038/s44222-023-00037-5

Stadler, L. J. (1928). Mutations of barley induced by X-rays and radium. *Science*, *68*(1756), 186–187. DOI: 10.1126/science.68.1756.186 PMID: 17774921

Stathakis, D., Savin, I., & Negre, T. (1994). Neuro-fuzzy modeling for crop yield prediction. *The International Archives of the Photogrammetry, Remote Sensing and Spatial Information Sciences*, *34*, 1–4.

Stavi, I., Thevs, N., & Priori, S. (2021). Soil salinity and sodicity in drylands: A review of causes, effects, monitoring and restoration measures. *Frontiers in Environmental Science*, *9*, 712831. DOI: 10.3389/fenvs.2021.712831

Stewart, C. N.Jr. (2025a). *Plant biotechnology and genetics: Principles, techniques, and applications*. John Wiley & Sons.

Stocker, T. F., Qin, D., Plattner, G. K., Alexander, L. V., Allen, S. K., Bindoff, N. L., Bréon, F. M., Church, J. A., Cubasch, U., Emori, S., & Forster, P. (2013). Technical summary. In *Climate change 2013: the physical science basis. Contribution of Working Group I to the Fifth Assessment Report of the Intergovernmental Panel on Climate Change* (pp. 33–115). Cambridge University Press.

Subash, N., Gangwar, B., Singh, R., & Sikka, A. K. (2015). Identification of climate-resilient integrated nutrient management practices for rice-rice cropping system—An empirical approach to uphold food security. *International Journal of Biometeorology*, *59*(1), 65–78. DOI: 10.1007/s00484-014-0825-5

Surendran, U., Raja, P., Jayakumar, M., & Subramoniam, S. R. (2021). Use of efficient water saving techniques for production of rice in India under climate change scenario: A critical review. *Journal of Cleaner Production*, *309*, 127272. DOI: 10.1016/j.jclepro.2021.127272

Su, W. H. (2020). Advanced Machine Learning in Point Spectroscopy, RGB- and Hyperspectral-Imaging for Automatic Discriminations of Crops and Weeds: A Review. *Smart Cities*, *3*(3), 767–792. DOI: 10.3390/smartcities3030039

Suykens, J. A. K., Van Gestel, T., De Brabanter, J., De Moor, B., & Vandewalle, J. (2002). *Least Squares Support VectorMachines*. World Scientific. DOI: 10.1142/5089

Taheripour, F., Hertel, T. W., Gopalakrishnan, B. N., Sahin, S., & Escurra, J. J. (2015). Agricultural production, irrigation, climate change, and water scarcity in India.

Talanova, L. A. (2006). Justification of the effectiveness of methods for treating cucumber seeds and plants with magnetized water and humic acids. Abstract of the dissertation for the degree of candidate of agricultural sciences in the specialty 06.01.06, Vegetable growing. Ryazan, 27.

Tanaka, A., & Navasero, S. A. (1966). Growth of the rice plant on acid sulfate soils. *Soil Science and Plant Nutrition*, *12*(3), 23–30. DOI: 10.1080/00380768.1966.10431192

Tasty and productive Lugovskoy potatoes. URL: https://rusfermer.net/ogorod/korneplody/kartofel/sorta/srednerannie/lugovskoj.html. Date of access 08/27/24.

Tege, S., Jangeed, D., & Soni, B. (2025). Exploring the role of digital technologies in promoting circular economy practices in the Indian agriculture sector. In *Water Sustainability and Hydrological Extremes* (pp. 333–357). Elsevier. DOI: 10.1016/B978-0-443-21499-8.00017-9

Thampatti, K. C., Cherian, S., & Iyer, M. S. (1998). Influence of liming materials on soil acidity characteristics of an acid sulphate soil. *Journal of the Indian Society of Soil Science*, *46*(2), 296–299.

Thayer, A., Vargas, A., Castellanos, A., Lafon, C., McCarl, B., Roelke, D., Winemiller, K., & Lacher, T. (2020). Integrating Agriculture and ecosystems to find Suitable Adaptations to Climate Change. *Climate (Basel)*, *8*(10), 10. DOI: 10.3390/cli8010010

Thiyageshwari, S., Gayathri, P., Krishnamoorthy, R., Anandham, R., & Paul, D. (2018). Exploration of rice husk compost as an alternate organic manure to enhance the productivity of blackgram in typic haplustalf and typic rhodustalf. *International Journal of Environmental Research and Public Health*, *15*(2), 358. DOI: 10.3390/ijerph15020358

Tretyakov, H.N. (2005). Physiology and biochemistry of agricultural plants. *Adaptation and stability of plants.* Moscow, Kolos, 640.

Tyagi, N. K. (2003). Managing saline and alkaline water for higher productivity. JW, Kijne, R., Barker, D., Molden (eds.). Water productivity in agriculture: limits and opportunities for improvement. CAB International, 69-87.

Tyczewska, A., Twardowski, T., & Woźniak-Gientka, E. (2023). Agricultural biotechnology for sustainable food security. *Trends in Biotechnology*, *41*(3), 331–341. DOI: 10.1016/j.tibtech.2022.12.013 PMID: 36710131

U.S. Salinity Laboratory Staff. (1954). Diagnosis and improvement of saline and alkali soils. In Richards, L. A. (Ed.), *USDA Agricultural Handbook 60*. USDA.

UNFCCC. 2017. Glossary of climate change acronyms and terms.

Uno, Y., Prasher, S. O., Lacroix, R., Goel, P. K., Karimi, Y., Viau, A., & Patel, R. M. (2005). Artificial Neural Networks to Predict Corn Yield from Compact Airborne Spectrographic Imager Data. *Computers and Electronics in Agriculture*, *47*(2), 149–161. DOI: 10.1016/j.compag.2004.11.014

Upadhyay, S. K. (2021). *Genome engineering for crop improvement*. Wiley Online Library. DOI: 10.1002/9781119672425

Van Klompenburg, T., Kassahun, A., & Catal, C. (2020). Crop yield prediction using machine learning: A systematic literature review. *Computers and Electronics in Agriculture*, *177*, 105709. DOI: 10.1016/j.compag.2020.105709

Vasiliev, V.V., Slesarenko V.V., Gurskaya, T.A., Primachev V.I. (2010). Activated water increases fruiting of vegetable crops. *Potatoes and vegetables*, 6, 27.

Veenadhari, S., Mishra, B., & Singh, C. D. (2011). Soybean Productivity Modelling using Decision Tree Algorithms. *International Journal of Computer Applications*, *27*(7), 975–8887. DOI: 10.5120/3314-4549

Verge, X. P. C., De Kimpe, C., & Desjardins, R. L. (2007). Agricultural production, greenhouse gas emissions and mitigation potential. *Agricultural and Forest Meteorology*, *142*(2-4), 255–269. DOI: 10.1016/j.agrformet.2006.06.011

Verma, K. K., Song, X.-P., Budeguer, F., Nikpay, A., Enrique, R., Singh, M., Zhang, B.-Q., Wu, J.-M., & Li, Y.-R. (2022). Genetic engineering: An efficient approach to mitigating biotic and abiotic stresses in sugarcane cultivation. *Plant Signaling & Behavior*, *17*(1), 2108253. DOI: 10.1080/15592324.2022.2108253 PMID: 35959678

Vivekanandan, E., Rajagopalan, M., & Pillai, N. G. K. (2009). Recent trends in sea surface temperature and its impact on oil sardine.

Wasti, S., Sah, N., & Mishra, B. (2020). Impact of heat stress on poultry health and performances and potential mitigation strategies. *Animals (Basel)*, *10*(8), 1266. DOI: 10.3390/ani10081266

Wu, W., & Ma, B. (2015). Integrated nutrient management (INM) for sustaining crop productivity and reducing environmental impact: A review. *The Science of the Total Environment*, *512*, 415–427. DOI: 10.1016/j.scitotenv.2014.12.101

Xu, Q., Zhang, T., Niu, Y., Mukherjee, S., Abou-Elwafa, S. F., Nguyen, N. S. H., Al Aboud, N. M., Wang, Y., Pu, M., Zhang, Y., Tran, H. T., Almazroui, M., Hooda, P. S., Bolan, N. S., Rinklebe, J., & Shaheen, S. M. (2024). A comprehensive review on agricultural waste utilization through sustainable conversion techniques, with a focus on the additives effect on the fate of phosphorus and toxic elements during composting process. *The Science of the Total Environment*, *942*, 173567. DOI: 10.1016/j.scitotenv.2024.173567 PMID: 38848918

Yadav, J. S. P., & Girdhar, I. K. (1977). Salt translocation and crop growth under sprinkler and surface methods of irrigation in sodic soil [India]. *Indian Journal of Agricultural Sciences*, •••, 47.

Yan, Y., Zhu, X., Yu, Y., Li, C., Zhang, Z., & Wang, F. (2022). Nanotechnology strategies for plant genetic engineering. *Advanced Materials*, *34*(7), 2106945. DOI: 10.1002/adma.202106945 PMID: 34699644

Yuvaraj, M., Bose, K. S. C., Elavarasi, P., & Tawfik, E. (2021). Soil salinity and its management. Soil moisture importance, 1, 109.

Yvoz, S., Petit, S., Biju-Duval, L., & Cordeau, S. (2020). A framework to type crop management strategies within a production situation to improve the comprehension of weed communities. *European Journal of Agronomy*, *115*, 126009. DOI: 10.1016/j.eja.2020.126009

Zecca, F. (2019). The Use of the Internet of Things for the Sustainability of the Agricultural Sector: The Case of Climate Smart Agriculture. *Int. J. Civ. Eng. Technol.*, *10*, 494–501.

Zelepukhin, I.D., Pasko O.A., Aksenova E.A. (1988). *Application of activated water in agriculture and biotechnology*. Tomsk: Science and production.

Zhang, D., Xu, F., Wang, F., Le, L., & Pu, L. (2025). Synthetic biology and artificial intelligence in crop improvement. *Plant Communications*, *6*(2), 101220. DOI: 10.1016/j.xplc.2024.101220 PMID: 39668563

Zhang, F., Cui, Z., Chen, X., Ju, X., Shen, J., Chen, Q., Liu, X., Zhang, W., Mi, G., Fan, M., & Jiang, R. (2012). Integrated nutrient management for food security and environmental quality in China. *Advances in Agronomy*, *116*, 1–40. DOI: 10.1016/B978-0-12-394277-7.00001-4

Zhang, S., Huang, W., Huang, Y., & Zhang, C. (2020). Plant species recognition methods using leaf image: Overview. *Neurocomputing*, *408*, 246–272. DOI: 10.1016/j.neucom.2019.09.113

Zhao, Q., & Dixon, R. A. (2014). Altering the cell wall and its impact on plant disease: From forage to bioenergy. *Annual Review of Phytopathology*, *52*(1), 69–91. DOI: 10.1146/annurev-phyto-082712-102237

Zheng, A., & Casari, A. (2018). *Feature Engineering for Machine Learning: Principles and Techniques for Data Scientists*. O'Reilly Media, Inc.

Zia, H., Rehman, A., Harris, N. R., Fatima, S., & Khurram, M. (2021). An experimental comparison of IoT-based and traditional irrigation scheduling on a flood-irrigated subtropical lemon farm. *Sensors (Basel)*, *21*(4175), 3–17. DOI: 10.3390/s21124175 PMID: 34204584

About the Contributors

Janardan Singh did his B.Sc. Ag. (1986) and M.Sc. (1988) from ANDUA&T, Kumarganj, Ayodhya and Ph. D. (1992) from CSK HPKV, Palampur. Dr Singh has served as Assistant Agronomist at CSK HPKV, Palampur (1992), Scientist at Potash Research Institute of India, Gurgaon (1992-1995), Assistant Extension Specialist (1995-2001) and Agronomist (2001-2006) at CSK HPKV, Palampur, Associate Professor at HAC, Keren, Eritrea (2006-2008) and Associate Director, CSKHPKV Regional Research Station, Kukumseri, Lahaul & Spiti (2010-2014). He has also served as Nodal Officer of National Academic Depository (NAD) Cell of the University from 2016 to 2020.Dr. Singh is presently serving as Head, Department of Organic Agriculture & Natural Farming, College of Agriculture, CSK Himachal Pradesh Krishi Vishvavidyalaya, Palampur. He is also President of Oraganic Agricultural Society of India and H P Agricultural University Teachers Association, Palampur. Dr Singh has professional experience of more than 32 years as Assistant Professor/ Associate Professor/Professor. He has made outstanding contribution in the refinement and development of crop production technologies besides transfer of technologies among the hill farmers. Dr. Singh has guided9 M.Sc. and 3Ph.D students as major advisor. He has completed 15 research/extension projects as PI/Co-PI. Dr. Singh has been awarded with 3 International and 15 National awards by International agencies/ ICAR/SAUs/Professional Societies. He has published 115 research/ conference papers, 3 books, 9 booklets, 3 book chapters, 43 manuals, 60popular articles, 5 brochures and 6 pamphlets/leaflets. Dr Singh has organized many trainings/refresher courses/Radio School/National

Conference as Coordinator/Training Organizer besides aNational Training/ Summer School as Course Director/Co-Course Director. He has visited countries like Israel as a trainee and Eritrea as a faculty member in HAC, KerenEriteria.

Shilpa did Ph. D. Agronomy (2022) from CSK HPKV, Palampur. She is ASRB-ICAR-NET qualified and the overall university rank holder during graduation and doctorate programme. Dr Shilpa has served as Assistant Professor at Abhilashi University,Mandi, Himachal Pradesh (2018) and Project Associate at CSKHPKV,Palampur (2022-23).Dr. Shilpa is presently serving as Assistant Professor, School of Agriculture, Lovely Professional University, Phagwara, Punjab. She has also been serve the responsibility of secretary (2020-21) and president (2021-2022) of Student Association (P.G.) at CSKHPKV, Palampur. Dr Shilpa has guided 4 M.Sc students as major advisor. She has been awarded with 2 National awards by Professional Societies. She has published 55 research/conference papers, one book, 26 book chapters and 13 popular articles (Hindi &English).Dr. Shilpa has attended 7 national trainings till date.

Pooja did B.Sc. Hons. Ag. (2017) and M.Sc. Agronomy (2019) from MPUAT, Udaipur and Ph. D. Agronomy (2023) from CSK HPKV, Palampur. She is ASRB-ICAR-NET and UGC NET qualified with 70.67% and 60%. She has published 12 research/conference papers, 10 book chapters & popular articles and attended 9 national trainings till date.

***

Satheesh C.S. received an B.E(EEE) degree from Mahendra Engineering College, affiliated to Anna University, during 2014-2018,India, M.E (Control Systems) degree from Mahendra Engineering College, Namakkal, Tamil Nadu, India in 2020. His research interests are Control Systems, renewable energy conversion, and soft computing techniques. At present he is a Assistant Professor in EEE department at Muthayammal Engineering College (Autonomous), Rasipuram, India. He has 4 years of teaching experience. His research findings have been published in SCIE/SCOPUS/Google-indexed international journals.

He has attended and presented many research articles at international conferences.

Gowrishankar J. has completed B.Tech in Information Technology and M.E. in Software Engineering and Ph.D. in Information and Communication Engineering from Anna University. I have 16.9 years of teaching experience. I have published many papers and book chapters in National and International journals as well as published 4 Patents in National level. My areas of interests are Internet of Things, AI, Networking and Cloud Computing etc., I handled the subjects like Internet of Things, Python for Data Science, Cyber Security and Risk Management, IT Infrastructure and Technologies, Cloud Computing, Software Engineering, Design and Analysis of Algorithms, Data Structures, Python Programming, OOPS using C++, Object Oriented Analysis and Design for UG and PG Students. Now Presently acting as an PCL & Project Coordinator for the AI department. Also acting as a Supervisor for a Ph.D. Research Scholars as well as handling Internet of Things core subject for the Research Scholars of CSE Specialization in JAIN University

Radhika K. is the Professor in the Department of Electronics and Communication Engineering at Muthayammal Engineering College. She has around 15 years of Teaching experience. She has published papers in reputed International Journals, Books/Chapters and published Patents. She is the life members of ISTE, IETE. Her area of interest includes VLSI, Embedded Systems and Instrumentation

Nadezhda Anatolievna Lebedeva is a Doctor of Philosophy in the field of Cultural Studies and works at the Kherson State Agrarian University (Kherson, Ukraine) as the department head of Humanitarian Disciplines Chair, and International Personnel Academy (Kyiv, Ukraine), professor, and a full member of the Eurasian Academy of Television and Radio. She is the main editor of the international journal "Universum: Philology & Arts", a reviewer and member of the editorial board of scientific collections of international conferences and scientific journals ("Journal of Social, Humanity, and Education", "Journal of Sustainable Tourism and Entrepreneurship" Goodwood Publishing), the author of more than a hundred publications, including Scopus and WoS, has recommendations and certificates of honour for successful scientific activity). Experienced in project cinematography management and developed lecture courses in a foreign language for students of state administration. She is the author of more than 100 scientific works. The focus of her research is the cultural and social aspects of regional development. N. Lebedeva graduated from the Kherson State Pedagogical Institute in 1996. After graduating from the post-graduate course at the Kherson State University, became a lecturer at the History of World Culture and Literature Department of the Kherson State University. In June 2008 was enrolled in the doctoral program of the Interregional Academy of Personnel Management in Kyiv, 556 specializing in Cultural Studies. She was awarded the medal "For Civil Courage" of the Afghan Veterans' Ukrainian Union for the book "My Afghan" by Nikolay Vishnyak". Participation in the creation of documentary films and scientific articles is marked by membership in the Eurasian Academy of Television and Radio (November 2012). The monograph "Cultural Aspect of Alexander Elagin's Creativeness" was awarded 1st place in scientific direction "Cultural Studies" in 2017 and became the winner of the competition for educational projects "Interclover – 2017" in the nomination – "Research Project".

Maheswari M. has awarded Ph.D. degree under Electrical Engineering from Anna University, Chennai. She is currently working as Senior Lecturer in the University of Technology and Applied Sciences, Nizwa, Sultanate of Oman. She has got 20 +years of experience in Academics and Administration. She has published more than 40 research papers in International journals, National journals, and International and National conference proceedings, 02 patents and 6 Book Chapters. She is the Life Member in ISTE and Member in IE(I). Her research interests include Renewable Energy applications in Power System Engineering.

Olga Anatolievna Pasko worked at the National Research Tomsk Polytechnic University, Engineering School of Natural Resources, Department of Geology, Professor. Currently, she is Vice-Rector for Research at the National Open Institute St. Petersburg. She has work experience in the "Agrophysical Research Institute", St. Petersburg as the head of the experimental research sector. Main responsibilities: management of profiles in the areas of "Land Management" (master's degree) and "Land management, cadastre and land monitoring" (postgraduate studies); preparation and implementation of accreditation for these specialities in 2016 and 2021; planning, organization and control of educational, educational and methodological work in the disciplines provided, taking into account the specifics of the contingent of students. Olga Anatolievna created some working curricula for the provided disciplines and management of their development by other teachers. She also conducted all types of training sessions, managing course and diploma projects and research work for bachelor's and masters. She is also known for her teaching experience at universities in Tomsk: National Research Tomsk Polytechnic University, National Research Tomsk State University, Tomsk University of Architecture and Civil Engineering, Tomsk Agricultural Institute – a branch of the Novosibirsk Agrarian University. More than 200 publications have been published, including in journals included in the Scopus and WoS databases. Awarded with thanks from the Ministry of Science and Higher Education of the Russian Federation, the Administration of the Tomsk Region, State Duma of the Tomsk Region, Administration of the Novosibirsk Region, Government of the Novosibirsk Region, Omsk State University, Tomsk State University, Tomsk Polytechnic University.

G. Sudha currently working as Professor & HOD / Biomedical Engineering in Muthayammal Engineering College. Received Ph.D from Anna University Chennai, in Electrical Engineering in the year 2018. M.E. in Process Control & Instrumentation from Annamalai University in 2005. B.E. in Electronics & Instrumentation from Bharathiyar University in 2003. Research Interests are Control System, Internet of Things, Soft Computing. Published 10 papers in International Journals, Presented 10 papers national & International Conferences. Obtained 5 Indian Patent and applied 3 Indian Patent. Reviewer for the peer – reviewed journal Advances in Science, Technology and Engineering Systems Journal. Resource person for various technical talks. Organized several workshops, technical talks, guest lectures, workshops

N. Suthanthira Vanitha has awarded Ph.D. degree under Information and Communication Engineering in the specialization of Bio-Medical and Embedded Systems from Anna University, Chennai . Currently working as a Professor in the Department of Electrical and Electronics Engineering and Dean-Quality Assurance at Muthayammal Engineering College, Rasipuram. She has 22 years of experience in Academic, Research and Administration. She has published 182 papers in National/International Journals and Conferences,12 Books/Chapters, 13 Patents and 25 Awards. She is the Life Member in IEEE, ISTE and Fellow Member in IE(I). Her area of specialization includes, Embedded Systems, Biomedical Instrumentation and Renewable Energy Systems.

Ravi Verma is working as Assistant Professor (Agronomy) in the School of Agriculture, Lovely Professional University, Phagwara, Punjab (ICAR Accredited & NAAC A++). He has completed his B.Sc. (Ag) & M.Sc. in Agronomy from C.C.R.P.G. College, Muzaffarnagar (NAAC 'B'), affiliated with Chaudhary Charan Singh University, Meerut (NAAC 'A++') and completed his Ph.D. in Agronomy from Acharya Narendra Deva University of Agriculture & Technology Kumarganj, Ayodhya (U.P.). (NAAC 'A++') in 2023. Dr. Verma has published 11 research papers, 13 book chapters, 26 articles in various popular magazine's and also attended many training programs, national and international conferences and seminar. He has awarded with best research scholar award in 2023 from NADCL, Jammu Kashmir and Agricultural Research Council of Nigeria, Nigeria

Index

R

S

W

Y

www.ingramcontent.com/pod-product-compliance
Lightning Source LLC
LaVergne TN
LVHW081525300926
849161LV00006B/1306